CASTELLO

INVESTIGATING
MATHEMATICS

STUDENT RESOURCE BOOK

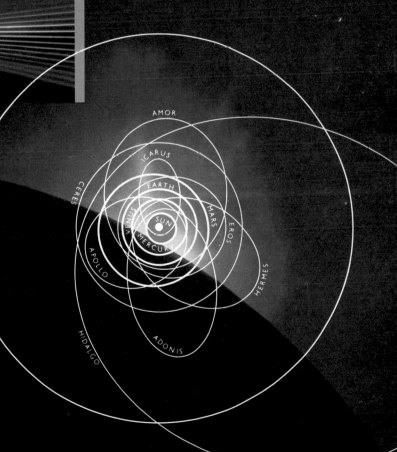

GLENCOE

Macmillan/McGraw-Hill

New York, New York   Columbus, Ohio   Mission Hills, California   Peoria, Illinois

Send all inquiries to:
Glencoe Division, Macmillan/McGraw-Hill
936 Eastwind Drive
Westerville, Ohio 43081

ISBN: 0-675-13160-X (Student Resource Book)

1 2 3 4 5 6 7 8 9 10 VH 02 01 00 99 98 97 96 95 94 93

DEVELOPMENT OF

INVESTIGATING

MATHEMATICS

An Interactive Approach

Investigating Mathematics: An Interactive Approach is the result of a unique publisher/state department of education partnership to publish innovative mathematics learning materials, known informally as "Math A." Beginning in 1988 and continuing through 1992, Math A was developed under the direction of Thomas J. Lester of the California Department of Education. Mr. Lester is the director of both the Math A and Math B Committees that seek to link local efforts to support and clarify the Math A/B concept. The Committees are composed of over fifty mathematics specialists, high school mathematics teachers, and college professors of mathematics who wrote, piloted, critiqued the initial materials, and then trained teachers throughout the state in using the materials. Without the hard work and dedication of the Math A Committee, the development of this program would not have been possible. For over three years, these materials were field tested in hundreds of schools by hundreds of teachers and thousands of students. Based on their use in schools, the materials were revised continuously.

The California Department of Education then requested for a publisher to join them in a partnership to publish the Math A. In the bid process, Glencoe Publishing Company was selected. Personnel from both the state department and Glencoe read writing samples submitted by potential authors to rewrite the materials. Dr. Larry Hatfield was chosen as author. In late 1991, Glencoe, the California Education Department, and Dr. Hatfield entered into a unique partnership to publish Math A. The result of this collaboration is *Investigating Mathematics: An Interactive Approach*.

DEVELOPMENT TEAM

Author
Larry L. Hatfield

Larry L. Hatfield is a Professor of Mathematics Education at the University of Georgia in Athens. Dr. Hatfield is Director of Project LITMUS, a five-year National Science Foundation supported teacher enhancement project to infuse computing technologies in the teaching of mathematics in grades K through 12 in two rural Georgia school systems. He received the Josiah T. Meigs Award for Excellence in Teaching, the University of Georgia's highest recognition for teaching. Dr. Hatfield received his B.S. and Ph.D. from the University of Minnesota and his M.A. from Western Michigan University. He has taught students and teachers at all levels of mathematics and has published in numerous journals and textbooks. Dr. Hatfield served at the National Science Foundation and has been active at conferences and with committees nationally and internationally.

Course Development Director
Thomas J. Lester

Thomas J. Lester is a Mathematics Education Consultant with the California Department of Education. Mr. Lester was formerly the K-12 curriculum specialist for mathematics and computers for the San Juan Unified school district. He has taught mathematics at the high school, middle school, and elementary levels and was co-director of a NSF grant for teacher enhancement called "Problem Solving Using Computers." He currently teaches part-time in the Mathematics Department of California State University at Sacramento. For the past three years, Mr. Lester has been the executive producer and/or producer as well as on-camera talent for more than forty one-hour live television programs that deal with topics such as the implementation of the California Mathematics Framework and mathematics that parents can use to help their children. He is also the co-author of several mathematics texts.

Contributing Authors

The Contributing Authors worked closely with the Development Team and Glencoe editorial staff as the Student Resource Book was finalized. Their contributions to the Student Resource Book, based on their extensive Math A classroom experience, were invaluable in the creation of this program.

W. Karla Castello

Karla Castello is a teacher of mathematics at Yerba Buena High School in San Jose, California. She received her B.A. from San Jose State University. Ms. Castello has been a mentor teacher of mathematics since 1988 and also an in-service teacher for Math A units since 1989. She served as an author and consultant during the development of each component of *Investigating Mathematics: An Interactive Approach*. In particular, her contributions were instrumental in the development of the Teacher's Edition for which she served as author. She has been a speaker at many mathematics conferences and is also active in numerous professional organizations, including the National Council of Teachers of Mathematics.

Sandie Gilliam

Sandie Gilliam teaches mathematics at San Lorenzo Valley High School in Felton, California. She is a mentor teacher and instructor for the Monterey Bay Area Mathematics Project and has served as a consultant for the California Education Department as well as local school districts and county offices of education. She was a California semi-finalist for the United States Presidential Award for Excellence in Science and Mathematics Teaching. Ms. Gilliam received her B.A. from San Jose State University. She served as author and consultant during the development of each component of *Investigating Mathematics: An Interactive Approach*. In particular, her contributions were instrumental in the development of the Portfolio Builder for which she served as author. Ms. Gilliam is an active speaker at many mathematics conferences and in-services and is a member of the National Council of Teachers of Mathematics and many other professional organizations .

Original Authors

Each of the following mathematics educators were Original Authors of a portion of the original Math A program developed under the direction of the California Department of Education. In addition, each Original Author reviewed the manuscript developed by Dr. Hatfield.

Kathleen M. Borst
Mathematics Teacher
Anderson Valley High School
Boonville, California

Samuel O. Butscher, Ph. D.
Mathematics Department Head
San Francisco Unified School District
San Francisco, California

Dianne L. Camacho
Mathematics Teacher/
  Staff Development Mentor
Warren High School
Downey, California

W. Karla Castello
Mathematics Teacher
Yerba Buena High School
San Jose, California

Dolores Dean
Mathematics Department Chair/Teacher
Burroughs High School
Ridgecrest, California

Donna A. Gaarder
Mathematics Teacher/Site Coordinator
Mission High School
San Francisco, California

Sandie Gilliam
Mathematics Teacher
San Lorenzo Valley High School
Felton, California

Ana María Golán
Curriculum Specialist
Santa Ana Unified District
Santa Ana, California

Rowena F. Hacker
Mathematics Teacher
Trabuco Hills High School
Mission Viejo, California

Sylvia A. Huffman
Mathematics Teacher
Del Campo High School
Fair Oaks, California

Liane B. Jacob
Mathematics Teacher
Century High School
Santa Ana, California

Joann M. Kennelly
Mathematics Teacher
Hawthorne High School
Hawthorne, California

John D. Leonard
Mathematics Teacher
Los Alamitos High School
Los Alamitos, California

Carolyn E. Nelson
Curriculum Developer
Walnut Creek, California

Dianne Pors
Mathematics Coordinator
East Side Union High School District
San Jose, California

Jeanne K. Shimizu-Yost
Mathematics Teacher
San Juan High School
Citrus Heights, California

Harris S. Shultz
Professor of Mathematics
California State University
Fullerton, California

Sandra G. Sigal
Mathematics Teacher
Downey High School
Downey, California

Dick Stanley
Mathematics Consultant
University of California
Berkeley, California

Dorothy F. Wood
Mathematics Teacher
Tamiscal High School
Larkspur, California

Sheryl Yamada
Mathematics Teacher
Beverly Hills High School
Beverly Hills, California

Reviewers

Each of the following mathematics educators were Reviewers of the manuscript for all components of *Investigating Mathematics: An Interactive Approach.* This program reflects their suggestions.

John D. Leonard
Mathematics Teacher
Los Alamitos High School
Los Alamitos, California

Anita Wah
Mathematics Consultant
San Rafael, California

TABLE OF CONTENTS

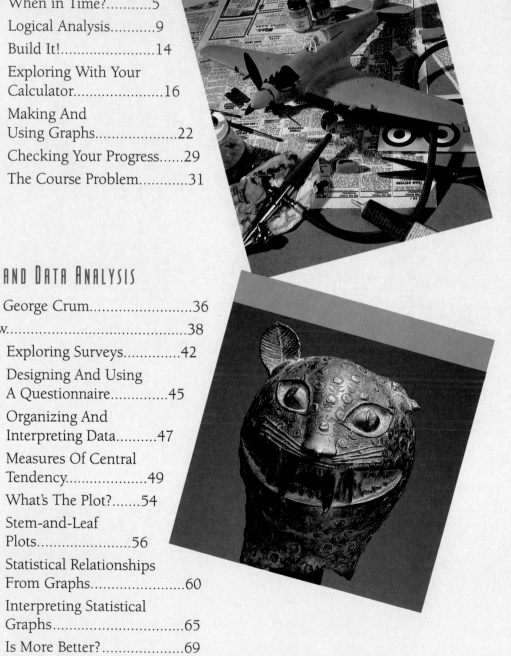

Unit 3: Relative Size of Numbers

Unit 4: Spatial Visualization

UNIT 5: WHAT ARE MY CHANCES?

UNIT 6: SIZE AND SHAPE IN THE PLANE

Unit 7: Thinking About Functions

Unit 8: Growth and Decay

COURSE OVERVIEW

ABOUT THE COURSE

This course has been designed to help you learn and enjoy mathematics. As the title of this book suggests, you will investigate many different situations in order to further develop your understanding of mathematical ideas and tools. An important part of this course is to help you learn how to plan and complete investigations using mathematics and to organize and document your findings about mathematics in a consistent, neat, and useful manner.

Using mathematics often requires us to explore as we solve a problem. Do you know how an explorer behaves? You will be asked to experiment with ideas in each situation. You will be encouraged to try out different ideas as you work to understand and solve the problems. This will require you to be actively involved rather than just receiving information and directions.

Mathematics helps us understand the many important issues and technical information in the world around us. As educated citizens, we need to know how to think and reason with mathematics. Such understanding can help us to respond to problems in our society and environment as well as in our everyday lives. An important goal of this course is to help you use mathematics to think about and help solve problems that concern us as citizens in our country and inhabitants of our planet.

WORKING IN GROUPS

One of the most important features of this course involves your work with other students. We all live and work in groups. It is important to learn to cooperate with others to complete a task. In almost every job in our society, people are expected to work together. Being able to cooperate with others is often vital for success in jobs and communities .

In this course, you will be expected to work and share with other students. In most of the activities, you will cooperate with others to complete a task.

To do the best possible job, each group member must cooperate as a responsible participant. Here are some suggestions for working effectively in groups:

- ♦ contribute your ideas positively
- ♦ listen to and appreciate the ideas of others
- ♦ show respect for others when you discuss and interact
- ♦ help each other to understand
- ♦ make decisions and predictions together
- ♦ be willing to revise your ideas
- ♦ help to obtain results and to write or give reports
- ♦ use your time wisely
- ♦ ask for outside help only when all group members have the same question
- ♦ do your share of the work
- ♦ show concern for the group success
- ♦ find the satisfaction of cooperating and enjoy your group activities

You may want to refer to this list of suggestions to check how well your group is working. Because your success in this course will largely depend on the progress of your group, you will want to be sure the group is working well. If it is not, your group should discuss why, and, if necessary, involve your teacher in helping to improve it.

As you can see, your role as a mathematics student will probably be different from any previous mathematics course you have taken. You can also see that the role of your mathematics

teacher will change. Your teacher will be less involved in presenting facts, giving directions, or demonstrating solved exercises. Instead, your teacher will guide and assist the problem-solving activities and investigations of groups and individuals. You will need to recognize the importance of and accept the responsibility for your own progress.

FEATURED PROBLEMS

One of the most important goals of this course is solving problems by applying what you have learned. You will use mathematics to develop solutions and to communicate results.

Each unit of this course will feature one or more Unit Problems to be solved. In addition, there will be an overall Course Problem to be solved. The Course Problem will be introduced near the end of Unit 1. It will be a global, general situation that may have definite solutions as well as solutions that can only be investigated and speculated about at this time. You may choose one of four Course Problem topics that include:

- ♦ Health   ♦ Wealth
- ♦ Housing   ♦ Population

The solutions that you find for these problems will involve many of the mathematical concepts and techniques introduced in this course.

The Unit Problems will be posed at the start of each of Units 2 through 8. At the time it is first stated, you will probably not be able to solve the Unit Problem, but you will discuss it in your group and you can begin to understand it. You will want to think about it and work on it as you progress through the activities of the unit. Exploring the problem and making a plan for how to solve the problem is just as important as finding the solution.

Because they are related, your work on the Unit Problems will help you to think about the Course Problem. In each unit, as you learn to reason with more powerful mathematical methods, you will be asked to think about the Course Problem and to make further progress in solving it. By the end of the course, you should be able to present a report of your work on the Course Problem.

A Word About Problem Solving

Solving interesting, challenging problems can be fun! Most people enjoy working on a puzzle or playing a game. Crossword or number puzzles, "who done it?" mysteries, soap operas, political issues, even our daily personal dilemmas are all situations that can challenge us. The satisfaction of finding a solution can feel very good. Some problems will take longer than others to solve, so don't give up if you can't find a solution right away. Try to be open to the fun of problem solving in mathematics .

With a positive attitude and by becoming more open to exploring situations together with other students, you will become a better problem solver. The first, and perhaps, most important goal in learning mathematics is to make sense of the mathematical ideas and tools. Because you will work in cooperative groups you will have the support of others as you share ideas and tasks. A special effort has been made to include many uses of mathematics so that you will see its relevance to the real world.

As You Begin

This is a new school year and a new mathematics course. This course has been designed to offer you unique experiences in mathematics. Of course, how successful these activities turn out to be will depend on you. Make a commitment to complete each problem. Be ready to make this the best year ever by giving it your best effort ever!

Starting Points

Nov. 20, 1923.

G. A. MORGAN

TRAFFIC SIGNAL

Filed Feb. 27.

2 She

Garrett A. Morgan (1877-1963) was born in 1877 in Paris, Kentucky. The seventh child in a family of eleven children, he was the son of a former slave. Morgan left school in the fifth grade and moved to Cleveland, Ohio. He was not to receive any further schooling. However, throughout his life, Morgan owned businesses that sold and repaired sewing machines, and he invented and manufactured products such as a hair straightener.

One day, Morgan witnessed an accident involving a car and a horse and carriage at an intersection. Three people were injured and the horse was injured so badly that it had to be shot. Morgan felt that something had to be done to prevent such accidents. In 1923, he devised and patented a three-way traffic signal that was widely used until the advent of electric traffic lights.

Garrett Morgan is also credited with inventing and patenting the first gas mask in 1914.

About This Unit

Mathematics is the study of patterns and relationships. We can find patterns all around us.

Patterns can be found in numbers. Number lines and graphs are important tools for organizing, ordering, and sequencing numerical data.

Patterns can also be found in the geometric shapes we see in the world. By using numbers to count or measure shapes we can find even more interesting relationships.

In this unit, you will see how a number line can be used to order and relate dates for historical events. You will study various graphs to see how they can be used to picture relationships. You will begin to work in a group to cooperate in solving geometric and numerical puzzles and problems. Since the calculator will be an important tool in this course, you will explore how to use a calculator to investigate mathematics.

WHEN IN TIME?

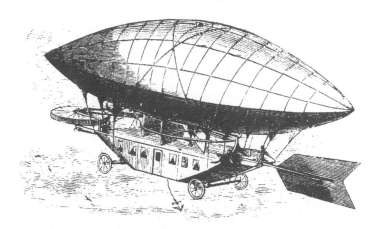

To help us understand any sequence of events, we can list them in the order in which they occurred. When events are listed on a numerical scale, we can more easily determine how much time passed between events. We can use such a **timeline** for history, literature, science, business, or even our family's heritage.

In this activity, you will use mathematics to make a timeline, a kind of number line that can be used to help you interpret past events. You will be part of a cooperative group to work and share with other students.

▶ PARTNER PROJECT

Create a timeline using events in aviation history.

1. Working with a partner, make a timeline for the aviation events listed.
   a. First, by yourself, study the list for a few minutes. Think about a plan for how you could organize the given information on a number line.
   b. Decide which dates will be assigned to the ends of the timeline. Determine the range of time for these dates. The **range** is the difference between the highest and lowest values. How will you place the dates so that there is equal spacing for equal units of time?
   c. Discuss your plan with your partner. Then, work together to construct a timeline as a team. Show each event with a point on the number line. Label each point.

AVIATION HISTORY	
EVENT	DATE
human balloon flight	1783
American jet	1942
scheduled passenger service	1919
Zeppelin flight	1900
solo trans-Atlantic flight	1927
Wright brothers' flight	1903
space shuttle	1981
first licensed woman pilot	1910
round-the-world non-stop flight	1949
rocket engine flight	1919
dirigible	1852
helicopter	1937
person in space	1961
airmail service	1918
trans-Atlantic flight	1919
man on the moon	1969

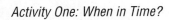

With your partner, use your aviation history timeline to answer the following questions.

2. Fold your timeline in half, and then in half again. In which quarter of the timeline do most of the events occur? Why do you think this is so? Explain why you think some of the events happened when they did.

3. Can you think of an aviation event that could have occurred earlier? more recently? Place these on your timeline.

4. How many years were there between the Wright brothers' flight and the first person in space? between scheduled passenger service and the American jet? Make another comparison that seems interesting to you.

5. Think about other ways that you could have displayed the information. How could you make a different kind of timeline using the same data?

HISTORICAL EVENTS

Discovery of the ozone hole	First reported case of AIDS
Start and end of the Black Plague	Stock Market Crash of the Great Depression
Exxon Valdez oil tanker spill	Start of the U.S. freeway system
World population reached 1 billion	Discovery of polio vaccine
First U.S. city to have 1 million people	First African-American in U.S. Congress
Completion of transcontinental railroad	Dupont chemical plant disaster in India
Soviet nuclear plant explosion	Spices brought from the Far East
Tomato introduced in Europe	Great famine in Ethiopia
Discovery of bacteria	

▶ GROUP PROJECT

Use the list of historical events above, or one developed by your class, to design and make a timeline.

6. With your partner, join another pair to make a group of four students.

   a. In your group, study the list of historical events. Decide who will research each event to find the date. Then use the date of each event to make a timeline.

   b. Use your timeline to find at least three interesting comparisons or relationships involving time. Present your ideas to the class.

7. Think of a kind of timeline you or your group would like to make. Some suggested topics are shown below. Develop a plan to decide who does what and the kind of events you want to include. Research your topic to find the information you will need to make your timeline. Make the timeline. Write a short report, using your timeline to help explain the ideas discussed in your report. Be sure to include a list of your resources.

   ◆ the U.S. space program
   ◆ major events in the history of your state or city
   ◆ history of a sport (for example: baseball, football, basketball)
   ◆ development of the music recording industry
   ◆ growth and development of a living animal or plant (for example: humans, cats, beans)
   ◆ history of the country where one of your parents or ancestors was born

8. COMPUTER ACTIVITY Study the following BASIC program that could be used to find the number of years represented by each centimeter of a timeline that spans 100 years. To make a timeline, you need to compute such a number.

   a. Write the expression in line 30 to complete the program. RUN the program to check results.

10 REM Timeline (years per cm)

20 INPUT "# OF CM IN TIMELINE?":length

30 LET yrspercm = _____

40 PRINT "# OF YRS PER CM = ";yrspercm

50 END

   b. Suppose we want to calculate how many years per centimeter for a timeline that spans any number of years. Add this line:

15 INPUT "# OF YEARS?":years
   and revise line 30. RUN the new program.

   c. Use your program to check the timelines you made in the previous exercises.

9. Make a list of situations where timelines or number lines would be used. Bring examples from newspapers or magazines.

▶ HOMEWORK PROJECT

10.  Read page viii of your Portfolio Builder. This page describes a portfolio and tells you the types of things you will be doing in your Portfolio Builder. Then, on page 1 of your Portfolio Builder, follow the directions to create a timeline to show the important events in your life. Share your timeline with your group or the class.

Choose either Exercise 11 or 12 to complete on your own. Be prepared to share your results with the class.

11. Geneology is the study of a family tree. Make a family timeline to show some important dates in your family, such as births or marriages.

12. Find out about the calendar of events for your school year. Make a timeline to show the school calendar.

LOGICAL ANALYSIS

Being able to solve mathematical problems depends on your ability to reason. Strategies help us to solve problems and win games that we play with our friends. Strategies also allow us to reason effectively in mathematics. In this activity, you will learn to play a game that will help you think about your own reasoning strategies. You will discuss your strategies with others in your group to find the sequence of fewest questions needed to solve the problem.

▶ PARTNER PROJECT A

Play *Rainbow Logic* to develop and discuss problem-solving skills.

Rainbow Logic is a game played on a square grid. Colored chips or tiles are used. For a 3-by-3 grid, there are three chips of each of three colors. The rules of play are as follows.

◆ One partner places the colored chips on a grid without showing the other partner. All of the chips of the same color must be on adjacent squares. Adjacent squares have at least one common side.

◆ The second partner takes a turn by selecting a row or column and asking what colors are in that row or column.

◆ The first partner tells the colors in that row or column, but not in any particular order.

◆ The goal is for the second partner to be able to tell the location of all of the colors on the grid in the least number of turns.

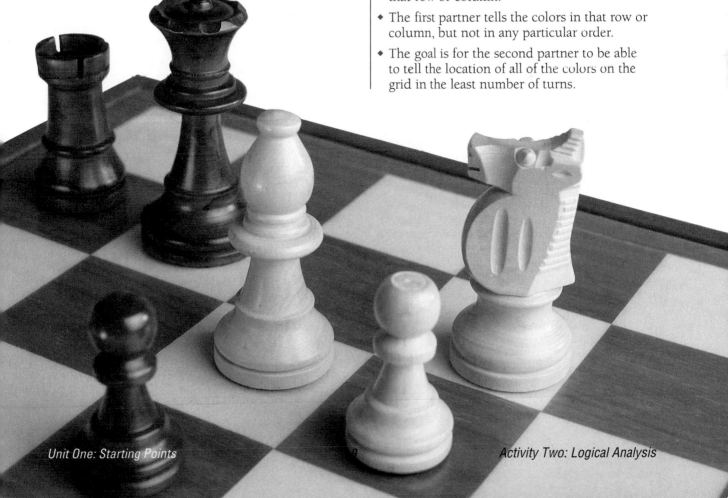

For example, colors may be hidden in a pattern like the one shown below.

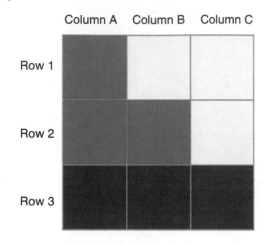

Column A   Column B   Column C

Row 1

Row 2

Row 3

Patterns like these are not allowed because squares of the same color do not share at least one side.

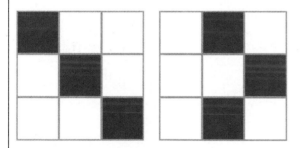

1. Play *Rainbow Logic* with your partner several times. Decide who will hide and who will find. After two games, exchange roles. As you play, think about how to select the next row or column, given what you may have learned from the previous turns. After each turn, discuss with your partner what you learn. Talk about a good way to keep track of what you learn from each turn.

2. After you have played several games, discuss with your partner any patterns you have noticed.

a. Write a brief summary of your game strategies on page 2 of your Portfolio Builder.

b. Show all of the different possible arrangements for three colors on a 3-by-3 grid on page 3 of your Portfolio Builder.

3. Play *Rainbow Logic* using four colors on a 4-by-4 grid. Again, play several games so each player is able to be the hider and the guesser. Try to use the strategies that you learned with the 3-by-3 grid. Compare the strategies used for a 3-by-3 grid with the strategies used for a 4-by-4 grid. What might be the fewest turns needed to know the locations of all colors for each type of grid? Be able to explain your reasoning.

4. Play *Rainbow Logic* on a 5-by-5 grid.
   a. Discuss with your partner what was the same and what was different about your sequence of questions when compared to the sequence with smaller grids. Your sequence of questions helps you identify your game strategy.
   b. Discuss other strategies that could have been used.
   c. What is the minimum number of guesses needed to find the locations of all colors on this grid?

d. Why do you think the number of guesses changed or stayed the same?

5. Play on a 3-by-3 grid with four colors. (Now it will be possible to have less than 3 squares of any color.) How does this change your strategy?

▶ GROUP PROJECT A

Use logic games to develop problem-solving skills.

Another **logic** game is called *Pico, Fermi, Bagels*. Rules for playing this game vary, but one way to play with two-digit numbers is explained below.

- In a group, one player (the leader) writes down a secret 2-digit number. These may include 00, 01, 02, and so on, through 99.
- Players take turns guessing different numbers.
- After each guess, the leader says one of these words:

Bagels	Neither digit is correct.
Pico	One digit is correct, but in the wrong place.
Pico-Pico	Both digits are correct, but in the wrong places.
Fermi	One digit is correct and in the correct place.
Fermi-Fermi	Both digits are correct and in the correct places.

6. Play the *Pico, Fermi, Bagels* game a few times with your group.
   a. After playing a few games, stop to discuss strategies that may reduce the number of guesses needed to find the correct number. For example, if you guessed 61 and were told "Bagels", what would you next guess? Why? What if you were told "Pico-Pico"? Why? If you guessed 57 and were told "Fermi", would you use these digits in your next guess? Why or why not?
   b. Play a few more games. Try to apply and improve your strategies. Keep track of the number of guesses for each game.
   c. Play the game with 3-digit numbers. Decide all of the responses the leader could make after each guess. Now try it with 4-digit numbers.
   d. Think about the strategies you used. Write the strategies you used for *Pico, Fermi, Bagels* on page 2 of your Portfolio Builder.

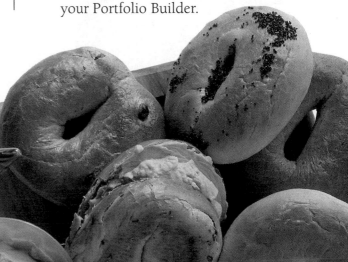

▶ PARTNER PROJECT B

7. Work with a partner to choose and study a strategy game. Some good ones include Mastermind®, Jotto®, Battleship®, Nim, Hangman, Tic-Tac-Toe, and 3-dimensional Tic-Tac-Toe.

   a. Analyze the game and develop your own playing strategy.

   b. Play the game with your partner and discuss your strategies.

   c. **PORTFOLIO BUILDER** Then write what you have learned about game strategies on page 2 of your Portfolio Builder.

▶ GROUP PROJECT B

Use Venn diagrams to organize information about membership in sets.

A **Venn diagram** uses overlapping circles to show which items are shared.

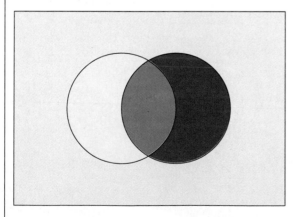

In your group, use the logic puzzle clues provided by your teacher to fill in a Venn diagram. Your group should have 15 people cards that can be placed in your Venn diagram.

8. Give at least one clue to each student in your group. Keep your clue hidden.

   a. Silently read your clue. As a group, decide how many different sets of items may be involved in all of the clues. Make a large two- or three-circle Venn diagram for group use.

   b. When it is your turn, read your clue aloud. As a group, use the clue to place the people cards in their appropriate place on the diagram. Check to be sure the condition of the clue is satisfied. Continue adjusting the people cards in your diagram as clues are read.

   c. **PORTFOLIO BUILDER** When all clues have been read, discuss the solution. Check the Venn diagram to be sure all conditions are satisfied. Record your solution on page 4 of your Portfolio Builder.

   d. **PORTFOLIO BUILDER** Think about the solution. Is it possible to show a different Venn diagram that still satisfies all of the clues? Record any alternate solutions on page 4 of your Portfolio Builder.

9. Discuss how your group found the items that belonged in each set. These questions may help your discussion.
   a. What made the puzzle difficult?
   b. Which clue was most helpful? Most difficult? Confusing or tricky?
   c. Which clue was the best one to start with? Why?
   d. Did you need all of the clues? If not, which clue(s) could have been left out?

10. **PORTFOLIO BUILDER** Think about the strategies you used. Write them on page 3 of your Portfolio Builder.

▶ GROUP PROJECT C

Construct a Venn diagram.

11. As a class, use recycling as a topic to create a Venn diagram. Label the parts of the Venn diagram.
    a. Working in groups of four, discuss your feelings about the selected topic. Respond individually by placing a sticky dot in the area of the Venn diagram that represents your feelings about the topic.
    b. Have a class discussion about the Venn diagram that was constructed. Discuss the meanings of the words *both*, *or*, *and,* and *not*. Use the Venn diagram to point out examples of each word.
    c. **PORTFOLIO BUILDER** Copy this class Venn diagram into your Portfolio Builder on page 5. Be sure to include the title and labels on your diagram.

BUILD IT!

Solving mathematical problems can be fun. It can be like solving a puzzle or a mystery. As you learn to use reasoning and logic more effectively, you will become a better problem solver.

In this activity, you will work in groups to solve some **spatial**, or **three-dimensional**, puzzles. By using a set of clues, your group will build a three-dimensional shape using colored cubes. Each clue will tell you something about the shape. You will decide which cubes to use from the clues.

▶ GROUP PROJECT

Use a set of clues to build a shape.

This project involves five *Build It!* puzzles with six sets of clues each. Work in your group to solve *Build It!* puzzle number 1.

1. For each set of clues, give at least one clue to each student in your group. Keep your clue hidden. Some sets of clues have special rules, which tell you the color of the cube(s) you can touch.

2. Start with some colored cubes. When it is your turn, read aloud your clue and use it to add a block to the shape. The person reading the clue is the only one that can touch the shape.

3. The clues may contain words that are unfamiliar to you, such as *edge* or *face*. Discuss in your group the meaning such words may have.

4. After your group has built a shape, check each clue carefully to be sure the shape satisfies all conditions. Sketch the shape and label the colors.

5. After all groups have finished, share your sketches or models with the entire class. How do they differ? Discuss how your group built the shape. These questions may help your discussion.

   a. What made the shape hard to build?

   b. Which clue was most helpful? most difficult? confusing or tricky?

   c. Which clue was the best one to start with? Why?

   d. Did you need all of the clues? If not, which clue(s) could have been left out?

   e. How did the special rules affect the solution?

6. Build puzzles 2 through 5. Choose one of your finished puzzles and record your solution on page 6 of your Portfolio Builder.

7. Think about the strategies you used. Write them on page 3 of your Portfolio Builder.

8. When we say that a solution is **unique**, that means that it is the only one.

   a. Do you think the shape you built for each set of clues is unique? Discuss why or why not.

   b. For each set of clues, try to build a different shape that satisfies all of the clues.

   c. Discuss how two shapes for the same clues are different.

▶ **PARTNER PROJECT**

Create a "Build It!" puzzle.

9. On your own, write your a set of "Build It!" clues. Include at least four clues in your set. Test your clues by building the shape to see if your clues produce a unique solution or the solution you predicted. Sketch your solution. Rewrite or add to your clues until your shape produces a unique solution.

10. Exchange your clues with your partner.

    a. Take turns and solve each other's puzzle. Make a sketch. Check to see if the solution is unique.

    b. Discuss how you each used the clues to find a shape. Observe how your partner reasoned with your clues.

EXPLORING WITH YOUR CALCULATOR

In today's world, it is important that you learn to use a calculator as a tool to help you solve problems. Calculators and computers are used by many people, including those working in business and industry. Most people use calculators in their everyday lives.

In this activity, you will explore the kinds of computational help **scientific calculators** provide. You will also see how a computer might be used. Throughout this course, you should use a scientific calculator to help you solve problems and explore mathematical concepts. A key part of learning to use a calculator is knowing whether you have used it correctly. Estimation will help you to see if your answer makes sense.

▶ PARTNER PROJECT A

Learn about some of the basic things your calculator can do.

1. Discover how you can use the operations on your calculator to make numbers.

   a. **PORTFOLIO BUILDER** Suppose you may use only 3s and any operation keys on your calculator. You may use no more than four 3s at a time. Produce all of the numbers from 1 to 20. For example, 3 ÷ 3 = 1. Record the keys you used for each one on the chart on page 7 of your Portfolio Builder.

   b. Now try to produce all of the numbers from 1 to 20 using any operation keys and only four of a number other than 3. (For example, four 5s). Record the keys used for each of these.

   c. Are there any cases where four of the same number didn't produce all of the numbers, 1 to 20? Explain your results.

   d. Study the charts you made. From these examples, what can you conclude about the **order of operations** on your calculator?

2. Explore the following examples to see how the **memory** on your calculator works.

Enter 7 in the memory by pressing the 7 key followed by the ⒮⒯⒪ key. An M will appear in the upper left corner of the display. The M means that there is a number stored in the memory.

a. Press the keys in the order indicated, and record your results on a piece of paper.

AC/ON	7	STO	___?___
	✕ 2	=	___?___
RCL	✕ 3	=	___?___
RCL	✕ 4	=	___?___
RCL	✕ 5	=	___?___
RCL	✕ 6	=	___?___

b. Look at the pattern formed by the numbers produced. What is this set of numbers?

c. Why don't you need to press RCL before ✕ 2? Why do you need to press RCL before each of the other multiplications?

d. What answers do you get if you don't press the RCL key before each multiplication? Explain what the RCL key does.

e. Repeat step 2a, starting with 9 rather than 7. What set of numbers are in this new sequence? Do you spot any patterns in these numbers?

3. Your calculator can help you find the **prime factorization** of a number.

a. Use the ⒮⒯⒪ and ⒭⒞⒧ keys to find the prime factorization of 9702.

b. Choose a year that was famous in history. Find the prime factorization of the numbers in that year using the ⒮⒯⒪ and ⒭⒞⒧ keys. How can you use the square root, √x̄, key to help you?

c. Describe the method you used on your calculator to find the prime factorization to your partner. Compare your methods.

d. 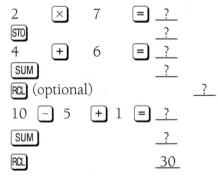 Complete the Sieve of Erasthosenes on page 8 of your Portfolio Builder.

4. The ⒮⒯⒪, ⒭⒞⒧, and ⒮⒰⒨ keys can help to do multiple step computations.

a. Record what is shown on the display after each step. Use this example to discover how these keys work.

2	✕	7	= ___?___
STO			___?___
4	+	6	= ___?___
SUM			___?___
RCL (optional)			___?___
10 − 5	+ 1	=	___?___
SUM			___?___
RCL			___30___

b. Why is the RCL step optional?

c. Write the arithmetic expression for this computation.

d. Use your calculator to check the computation for the expression below.

$(28.3 \times 7) + (173 + 16) + (312 - 42 + 7.8) = 664.9$.

5. Use the (STO), (SUM), and (RCL) keys to solve the following problem. Be sure you can explain the sequence of keys used to obtain the answer.

Suppose your class is selling food at your school's football games. How much will all of these items cost? Estimate the total cost, then use your calculator to find it.

- 125 packages of hot dogs at $1.79 a package
- 86 packages of buns at $0.89 a package
- 145 bags of chips at $1.69 a bag
- 94 2-liter bottles of soft drinks at $1.09 each
- 165 boxes of ice cream bars at $3.55 a box

6. COMPUTER ACTIVITY Study and complete the **spreadsheet** template that can be used to find the costs for the football concession above.

	A	B	C	D
1	Items	# of Items	Price of each	Cost of Items
2	hot dogs	125	$1.79	= B2*C2
3	buns	_____	_____	_____
4	chips	_____	_____	_____
5	soft drinks	_____	_____	_____
6	ice cream bars	_____	_____	_____
7	Total Cost			_____

Enter the values and expressions into a computer spreadsheet program. Use it to calculate the total cost. Find the total cost on your calculator and compare it with the result you found on your computer.

7. With the following examples, explore the way your calculator "remembers" the last operation used.

   a. Record the numbers for each blank.
   $\boxed{\text{AC/ON}}$ 6 $\boxed{\times}$ 10 $\boxed{=}$? $\boxed{=}$? $\boxed{=}$?

   b. Explain what is happening.

   c. Without clearing the display, try these.
   5 $\boxed{=}$?   8 $\boxed{=}$?   24.3 $\boxed{=}$?
   What is the constant now?
   8 $\boxed{+}$ $\boxed{=}$? $\boxed{=}$? $\boxed{=}$?
   How did the constant change? Why?
   5 $\boxed{+}$ $\boxed{=}$? $\boxed{=}$? $\boxed{=}$?

   d. $\boxed{\text{AC/ON}}$ 8 $\boxed{+}$ 100 $\boxed{=}$?
   Then 24 $\boxed{=}$? 5.6 $\boxed{=}$?
   $\boxed{\text{AC/ON}}$ 19 $\boxed{-}$ 5 $\boxed{=}$?
   Then 10 $\boxed{=}$? 56 $\boxed{=}$? 2 $\boxed{=}$?
   $\boxed{\text{AC/ON}}$ 24 $\boxed{\div}$ 2 $\boxed{=}$?
   Then 36 $\boxed{=}$? 25 $\boxed{=}$? 8.6 $\boxed{=}$?
   Explain in your own words how your calculator stores a constant. This is called the **constant function.**

8. Use the constant function to explore these sequences of numbers. In each case record the first ten numbers that result, tell what these numbers are, and make a number line to graph them.

   a. Count by 6 from zero.
   b. Multiply by 3 from one.
   c. Count down by 4 from 15.

   d. Divide by 10 from 1000.
   e. Divide by 2 from one.
   f. Divide by 5 from one.

9. Explore the fraction key, $\boxed{\text{a\%}}$, and the $\boxed{\text{d/c}}$ function (to get the $\boxed{\text{d/c}}$ function, press the 2nd and $\boxed{\text{a\%}}$ keys) on your calculator.

   a. Explore how to enter fractions or mixed numbers, and how to convert among decimals, mixed numbers, and improper fractions. Record the keys used.
   (1) Convert $2\frac{5}{8}$ to a decimal.
   (2) Convert $2\frac{5}{8}$ to an improper fraction.
   (3) Convert $\frac{34}{6}$ to a mixed number.

   b. Use your calculator to compute the following. Then, make a number line and graph each result.
   (1) $1\frac{1}{2} - 1\frac{1}{4}$
   (2) $\frac{3}{8} + \frac{5}{16}$
   (3) $\frac{1}{2} + \frac{7}{8}$
   (4) $\frac{1}{2} \div \frac{1}{8}$
   (5) $6 \times \frac{1}{3}$
   (6) $3\frac{7}{8} - 2\frac{1}{4}$

▶ **PARTNER PROJECT B**

Use your calculator to investigate mathematical patterns.

10. Use the reciprocal key, $\boxed{\text{1/x}}$, in the following investigations.

   a. For each of these numbers, find the **reciprocal.** Graph these numbers and their reciprocals on a number line. What can you conclude about the location of a number and its reciprocal?

$2, \frac{3}{4}, -\frac{1}{4}, \frac{1}{3}, \frac{5}{4}, -\frac{3}{2}, 2\frac{3}{4}$

b. Many patterns can be found in the digits of the decimals that represent sets of proper fractions. Use your calculator or a computer spreadsheet to complete the chart on page 9 of your Portfolio Builder. Describe the patterns you find.

c. Devise a method to find the decimals for $\frac{1}{17}, \frac{2}{17}, \ldots, \frac{16}{17}$ on your calculator. Describe any patterns you find in the digits of these decimals.

11. Suppose we call pairs of two-digit numbers, such as 51 and 15, 49 and 94, or 82 and 28, "cousins." We want to find out how far apart the cousins live. That is, we want to find the difference between a two-digit number and its cousin. Always subtract the lesser number from the greater number.

a. Try some examples. Look for patterns.

♦ Group the "cousins" that have the same difference. How are these various pairs alike?

♦ Make another group of "cousins" based on their differences being in a certain range of ten, such as the teens or twenties. How do their differences vary?

♦ Make a number line, graph cousins, and connect the points for "cousins." What does this show? Organize and record your results on the chart on page 10 of your Portfolio Builder.

♦ Use the chart to find a pattern. In your Portfolio Builder on page 10, write a rule that will work for this pattern. Test your rule on some other examples. Record your results in your Portfolio Builder.

b. Make a group of three-digit "cousins." In three-digit cousins, the second digit always stays the same, but the first and third digits are reversed. Test the rule you wrote in your Portfolio Builder on your three-digit "cousins." Does your rule still work? If not, write a new rule that does work.

12. Use your calculator to explore this number puzzle.

Step 1 Name a 3-digit number (using three different digits).

Step 2 Reverse the digits.

Step 3 Subtract the lesser from the greater (call this A).

Step 4 Reverse these digits (call this B).

Step 5 Add A to B (call this the result).

Try these steps with several 3-digit numbers.

a. What is the result? Explain why you think this result occurs.

b. Can you prove that this number always results?

c. Create a number puzzle to be solved by others using a calculator.

13. **PORTFOLIO BUILDER** Investigate the following situation. Keep a well-organized, complete record of your findings on pages 10 and 11 of your Portfolio Builder. We will refer to this situation in later units.

◆ Make two sets of six cards each, say red and black. In each set, write 1, 2, 3, 4, 5, and 6 on the cards.

◆ Randomly pair one red and one black card for all 12 cards. Find the six differences (subtracting the lesser from the greater).

◆ Find the product of the six differences. Your goal is to produce an **odd number.** Explore using your calculator.

a. What odd numbers (less than 100) can be found? Make a chart and record each set of six pairings for each odd number. What is the greatest? Least? Explain how you know.

b. For any of these odd numbers, how many ways can you pair the cards to make it? Explain.

c. Are there any odd numbers that cannot be produced as a product of the six differences? Discuss your findings.

d. Are there odd numbers that can be produced in more than one way? If yes, check to be sure you have different pairings. Try to find all possible ways for each odd number. Which odd number has the greatest number of different ways it can be made?

MAKING AND USING GRAPHS

One of the most powerful ways to represent and compare sets of numbers is in a graph. In Activity 1, you learned to make a timeline to show events. A timeline is an example of a **number line** with an **origin** (zero) and a unit (length of one) of measure. Each point represents some single value, and the set of points for the special dates is a graph.

In this activity, you will see how to make graphs from pairs of numbers, using two number lines. Each point will represent two values. Two-dimensional graphs show more clearly how two quantities vary or are related. Be sure to use your calculator as you investigate these situations.

▶ **PARTNER PROJECT A**

Interpret graphs of two quantities.

In a **rectangular,** or **Cartesian, coordinate system,** two perpendicular number lines, called **axes,** are used. The axes intersect at point zero for each axis. Each axis is like a ruler for measuring some quantity. Each point of a graph has a value found on the **horizontal** (left to right) axis and a value found on the **vertical** (up and down) axis. For each point there is an **ordered pair** of numbers.

1. Consider how the quantities age and height might be related. Study the graphs shown. Each point represents an age (in months), and a height (in centimeters). Where is the point for 0 on each axis?

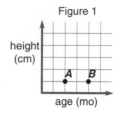
Figure 1

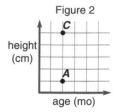
Figure 2

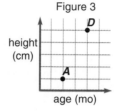
Figure 3

a. In each graph, which point represents the older person? Which is the taller person? How do you know?

b. In which graph are the ages the same? Explain.

c. In which graph are the heights the same? How do you know?

d. What does the graph in Figure 3 show?

e. Sketch a graph for two persons where the younger person is taller. How does it compare to the other graphs? Explain your reasons to your partner.

2. Record as ordered pairs the age (in months) and height (in cm) for you and your partner. Get together with another pair and record their information.

a. Study the data. What conclusions can you reach from the ordered pairs?

b. Sketch a graph that shows the correct location for each pair of numbers. Label the points. Why does a graph rather than a table or chart make it easier to understand how the several pairs of values may be related?

c. Which person is tallest? youngest? Which persons are closest in height? Are any students the same age? Which point represents the student who is the shortest and oldest?

d. Write a paragraph that describes the relationship between height and age found in the graph.

▶ HOMEWORK PROJECT A

3. **PORTFOLIO BUILDER** Refer to the graphs on pages 12 and 13 of your Portfolio Builder.

a. Find your age and height on the graph. Use the graph to make a prediction about your height at age 5, 10, and 20. Write the ordered pairs in a table. Explain how you made the predictions.

b. Is your growth a continuous process? Why or why not?

c. Compare the two graphs. How are they different?

▶ PARTNER PROJECT B

4. In your group, study the graphs. The horizontal axis in each graph measures time like a timeline. Try to match each description with its graph, and tell how the horizontal axis might be made. Various answers may be possible. Be ready to explain your reasons.

a.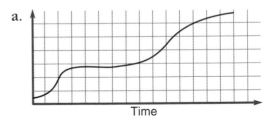
Time

b.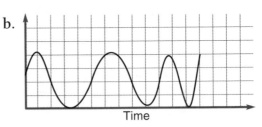
Time

c.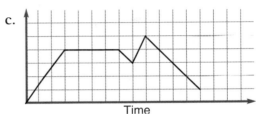
Time

d.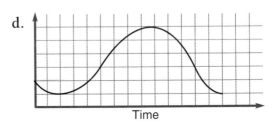
Time

- ◆ air temperature from midnight to evening
- ◆ driving a car up to speed limit, braking, passing, then stopping
- ◆ a person's growth record from childhood to adult
- ◆ height above the ground when riding on a seesaw
- ◆ the value of a house
- ◆ size of population of deer in the wilderness

5. Find additional examples of graphs that use a time as the horizontal axis. You may want to use newspapers or texts from your other courses, such as social studies or science. Make a display of the graphs.

6. Graphs can be used to show rates, such as speeds. A speed compares time and distance. Suppose you can ride your bicycle 1 kilometer in 2 minutes and 45 seconds. Each point of the graph is found from a pair of numbers, distance (kilometers) and time (minutes).

   a. How would a graph show this rate of speed? Where would the point be for 2 kilometers? For each additional kilometer, how does the time change? Make a graph to show your responses.

   b. Use the [a%] key on your calculator to enter a 1 kilometer time of 2 minutes 45 seconds. Use the [STO] key to "remember" this rate. Use your calculator to find how long it would take you to ride 12 kilometers at this rate. Find this point on the graph.

   c. Use the [RCL] key to help you find how long it would take to ride 15 kilometers. Where would this point be on the graph?

   d. Use the stored rate to find how many kilometers you could ride in 44 minutes. Plot this point.

   e. Use the stored rate to find how long it would take to ride 25 kilometers. Plot this point.

   f. Connect the points. What pattern do you see? Remember that you are riding a bicycle. What are some things that could happen while riding to change this pattern? Choose one thing that could happen to change the pattern. Describe this situation. Extend the graph, plotting what the graph would look like.

7. One kind of rate can involve cost per unit for a product or service. Study the points of the graph below which show costs per pound. Each of the labeled points represents some item.

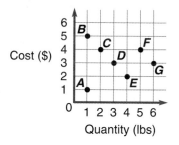

   a. Which item is the heaviest? Least expensive? Which items have the same weight? Same cost?

   b. Which items have the same value (for example, cost per pound)? What is the value? Connect the points for the same cost per pound. What appears to be true?

   c. Which item is the best buy or best value? Explain.

   d. Which item is the worst buy or worst value? Why?

   e. Find two items that lie on the same vertical line. Interpret what this means.

f. Find two items that lie on the same horizontal line. Interpret what this means.

▶ HOMEWORK PROJECT B

8. Choose a type of product.

a. Collect price information for your product so that you have at least eight different brands or sizes. Record the cost and the unit of measure for each item.

b. Make a graph of the ordered pairs (size, cost) for your product. Write a brief report about your data, using the questions posed in Exercise 7 to help guide your analysis.

c. Compare and discuss your graph and report with your class.

▶ GROUP PROJECT A

9. Study the following information.

- Cars A and B drive the same distance, but car B uses more gas.

- Car C uses the same amount of gas as does car D, but car C travels the same distance as car A.

- Car D travels farther than does car A.

- Cars E and F use the same amount of gas but car F travels farther than does car E, which travels the same distance as car D.

a. On a piece of paper, make a large graph based on the information given above. Display your graph.

b. Compare your graph with those made by other groups of students. How are they alike? Different? Explain.

c. As a class, discuss the graphs. Use these questions.

- Which car uses the greatest amount of gas? least? Which cars might use the same amount of gas? go the same distance?

- Which cars may have the same gas mileage?

- Which car appears to have the best gas mileage? Explain.

- Which car may have the worst mileage? Why?

10. Grocery stores abound with products that make fat percentage claims. You can find such products throughout the supermarket—-in the rows of salad dressings, in the packaged meat case, in the food freezer—-even in the bakery. How much fat is in your diet?

a. Look at the labels of several of your favorite foods. How many grams of fat are there in each serving? Use the total grams per serving to find the percentage of fat by weight. Use the percent key, %, on your calculator. Record your results in a chart.

b. It is recommended that no more than 30% of total calories each day should come from fat. There are about 9 calories per gram of fat. From the labels, find the percentage of fat by calories. Record these in your chart. Compare these values with those found in part a. Are any product labels misleading? Explain your results.

c. Find the fat content of the foods you eat in a typical day. List the foods you would eat, and find out how much fat might be included. Use your calculator to find the percentage of fat by weight and by calories. Does your diet meet the recommended 30% limit?

d. Do some package labels that tell the fat content of a food give misleading information? Think about words you see on labels such as *lite, fat free, lowfat, cholesterol free, nonfat,* and *extra light.* Find an example of a label that is misleading.

▶ HOMEWORK PROJECT C

Use graphs to solve problems.

11. Today lots of people are exercising for their health. To improve the condition of your heart, you need to work hard enough to benefit it, but not so hard that you overdo it. During exercise, you want your heart rate, the number of beats per minute, to be between your minimum aerobic heart rate and your maximum aerobic heart rate.

    a. Use the formulas below to estimate your minimum and maximum aerobic heart rates.

minimum aerobic heart rate = (220 – your age) × 0.69
maximum aerobic heart rate = (220 – your age) × 0.83

b. What is the range of your estimated heart rates?

c. Find the minimum and maximum values for members of your family or some of your friends. Get data from at least 10 people. Record the information in a chart.

d. What would be the age of someone whose maximum aerobic heart rate is 137? Whose minimum aerobic heart rate is 102?

e. Make a graph showing the minimum data. Above the minimum data, plot the maximum data. Can you use the graph to find the minimum and maximum heart rates for any age?

12. Tom is training to compete in a cross-country bicycle race.

   ◆ To warm up, Tom rides 25 kilometers per hour (km/h) for 20 minutes.

   ◆ Then he rides 30 km/h for 30 minutes, 35 km/h for 20 minutes, and 40 km/h for 10 minutes.

   ◆ To cool down, he rides for 30 minutes at 28 km/h.

   a. Make a chart to show the elapsed time and distance for Tom's ride.

   b. Use the values from the chart to make a graph showing time and total distance. How does the graph change as Tom's rate changes?

   c. Design a different training schedule for Tom. Make a chart and a graph to illustrate how the time and distance varies. Compare your training schedule with the one described above.

13. Another kind of rate involves the hourly wages of workers.

   a. Survey the members of your class to find those who have jobs that pay by the hour. Select four different hourly rates. Make a chart with at least six ordered pairs (number of hours,

wages earned) for each different hourly rate.

   b. Plot the ordered pairs from the charts on a graph. Connect the points.

   c. Compare the graphs. Which hourly rate is the greatest? Least? What is the shape of each graph? How are the graphs alike? Different? How could you use the graphs to find wages earned for examples you did not plot?

▶ **GROUP PROJECT B**

Use graphs to solve problems.

14. When people get together, they often shake hands with one another. Suppose everyone in your class decides to shake hands with everbody else in your class. How many handshakes do you think there would be?

   a. **PORTFOLIO BUILDER** First think about a simpler problem. Try to find out how many handshakes there would be in your group. Start with two people. Have one student in your group keep track of the number of students and the number of handshakes. Add another person each time until all group members have been added. Record each result in the table on page 14 of your Portfolio Builder.

   b. Now have your group join another group. Find out how many handshakes there would be in this larger group. Remember to add one group member at a time. Record your results in the table.

c. Study the data in the table. Find a pattern to predict the next result. Check it. Discuss how your groups got started. What problem-solving techniques did you try? What drawings, charts, or tables did you use? Did you find a formula or rule that helped you?

d. **PORTFOLIO BUILDER** Graph the data (number of people, number of handshakes) on page 14 of your Portfolio Builder. What can you conclude from the graph?

e. Use your calculator to extend the table to find how many handshakes there would be for the number of people in your class.

f. Even with your calculator, it would take a long time to make the table for hundreds of people. Look for a rule that would let you compute directly the handshakes for the number of people in your school.

15. COMPUTER ACTIVITY There are two simple ways to use the computer to help you find the number of handshakes. In each case refer to the data table from Exercise 14 above.

a. Complete this BASIC program to produce the sequence in your table for whatever number of people you input. You might find how many handshakes for all the people of your town or city.

```
10  REM Handshakes Problem
20  INPUT "How many?": people
30  LET handshakes = _____
40  FOR n = 1 TO _____
50  LET handshakes = _____+_____
60  NEXT n
70  PRINT "# of handshakes is";_____
80  END
```

b. Copy and complete the following spreadsheet to produce your table of values.

	A	B
1	No. of People	No. of Handshakes
2	2	1
3	= A2 + 1	= _____ + _____
4	= _____	= _____ + _____
5	= _____	= _____ + _____

Use the spreadsheet to find how many people are needed to produce at least 1 million handshakes.

CHECKING YOUR PROGRESS

As you continue in this course, you will be asked to approach how you learn mathematics in many new ways. You will explore mathematical situations to discover patterns and rules. You will solve real-life problems to see how mathematics can be useful. You will conduct investigations and develop projects, using various tools and reference materials. You will have the opportunity to work cooperatively with others as you discuss and solve problems. Above all, you should strive to make sense of mathematics as a way of thinking.

In most situations, it is important to assess your progress. As you and your teacher assess your progress in this activity, you will learn more about your strengths and weaknesses and you will have chances to reflect on them. This will enable you to feel proud of your work and help you focus on what needs to be improved. Assessments should be a positive, shared effort to understand your mathematical development.

PART I GROUP ASSESSMENT

Reflect on what you have learned in this unit.

In your group, identify the different topics that you have studied in this unit. List the topics and discuss them. Make a group poster showing what you have learned.

PART II INDIVIDUAL REFLECTION

Use the assessment sheets on pages 15 and 16 of your Portfolio Builder.

1. For each topic listed on page 15 of the assessment sheet in your Portfolio Builder, write a short paragraph explaining what you have learned about it. Include an example illustrating your understanding of each topic.

2. Use the three line segments on page 16 of your Portfolio Builder.
   a. Indicate your understanding of the topic by placing a letter for each topic on line segment 1 where appropriate.
   b. Indicate how much you liked a topic by placing a letter for each topic on line segment 2 where appropriate.
   c. Indicate how much you value what you have learned about the topic by placing a letter for each topic on line segment 3 where appropriate.

PART III PARTNER ASSESSMENT

Check your understanding by completing the following on pages 17 through 20 of your Portfolio Builder.

3. Make a number line and graph each number below.

   9   -2   3.4   -0.5 $\frac{4}{5}$
   $2\frac{3}{4}$ $\frac{19}{3}$ $\sqrt{13}$ $(2.6)^2$

4. Sketch a shape that uses 2 red cubes, 2 blue cubes, and 1 yellow cube. Be sure

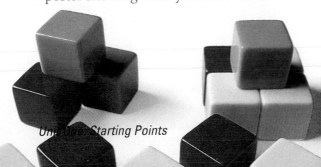

to label the colors on your sketch, or color them using crayons or markers. Write a set of clues that could be used to build your shape.

5. Make a graph to locate the cars, using the information given. Write a brief report telling three other facts you can find, using the graph.

   ◆ Cars A and B have the same speed but car B travels farther than car A.

   ◆ Car C travels the same time as car A but farther than car B.

   ◆ Car D travels farther and longer than cars B and C.

   ◆ Car E has a slower speed than car A but travels the same amount of time.

6. Use a calculator to find each of the following. Record the keystrokes that you used for each exercise.

   a. $\dfrac{(12.3 - 4.6)}{(1.2 \times 0.5)}$

   b. the square of 6.03

   c. the reciprocal of π

   d. the sum of 23×7, 23×11, and 23×0.7

e. the fraction and decimal for $3\frac{4}{7}$

f. the mixed number for 4.58

g. the greatest value among these prices
   i. $5 for 2 kg
   ii. $7 for 3 kg
   iii. $12 for 5 kg
   iv. $18 for 8 kg
   v. $22 for 10 kg

7. Use a calculator to find the first ten nonzero multiples of 6, 9, and 4. Make a Venn diagram to show the thirty numbers. What does the inner region of the circles represent?

8. Use a calculator to compute each expression and explain what order a calculator does operations and parentheses.

   a. $5 \times (4 + 2) - 18$

   b. $21.6 \div 4 - 0.6 \times 3$

   c. $4.3 - 2\frac{3}{4} \div \frac{5}{6}$

THE COURSE PROBLEM

There are many significant problems to be solved in our world today. One of the purposes of your education is to help you become a more concerned, knowledgeable citizen of the world. As such, you will want to know how mathematics might be used to understand and help to solve these important global problems.

Throughout this course you will investigate topics of social importance using mathematical ideas. Four topics, dealing with **housing, wealth, health,** and **population** are introduced in this activity.

You will choose one of these topics to investigate. Or you may choose to investigate all four topics with the idea that you would choose one of them for your final project. At some time during this course you will join a Course Problem Group that will work on the topic(s). At the end of the course, each course problem group will make a presentation about their topic to the class. To help you choose the topic that interests you, complete Group Project A.

▶ **GROUP PROJECT A**

Begin to understand the issues and questions for each Course Problem topic.

1. As a group, read the topic overviews beginning on page 32 for all four topics.

2. Discuss all four topics using the suggestions below.
   a. For each essay, what is the main focus or theme?
   b. Each of these topics involves many complicated problems to be solved in your world. Review the questions posed in each topic overview. What do you think is the most important question asked? Why?
   c. Write down your choice of topics in order from most favorite to least favorite.

▶ COURSE PROBLEM GROUP PROJECT A

3. For your topic(s), select one of the bulleted sections listed in the overview. How do you think you could use mathematics to think about those questions?

 a. For each topic, there are many more questions that relate to the bulleted section you have chosen. Brainstorm to make a list of additional questions that could relate to your topic(s).

 b. Narrow down the list of questions that interest your group. Choose one or more questions that you will research for your Course Problem. Your group may want to work on the same question(s) or each member may want to choose a different question within your topic(s). Decide what your group will do and the role of each member.

▶ COURSE PROBLEM GROUP PROJECT B

Prepare a timeline and a report for your Course Problem topic.

4. Make a list of key events for your topic.

 a. Find the dates when each event occurred. Make a list of the resources you used and which ones were the most helpful.

 b. Make a timeline to display the chronology of the events. Try to produce a neat, artistic timeline suitable for displaying in your classroom or hallways. Write a brief synopsis of each event which can be included on your timeline display.

▶ A FINAL WORD

Please remember that the Course Problem is a year-long endeavor. You should not expect to complete the Course Problem in a short time or with little effort. The timeline you will make in this unit, along with items you will make for the Unit Problems in Units 2 through 8 will help you with your course problem. At the end of each unit, your Course Problem Group will meet to see how the mathematics learned in the unit can be used in your Course Problem presentation.

COURSE PROBLEM TOPIC OVERVIEWS

HOUSING Having adequate shelter is a basic human need. One of the largest areas in the United States economy is the housing industry. And one of the largest portions of a family's budget can involve the costs of housing. Perhaps you have taken your own housing for granted. For many people, affordable housing is a major problem. Today in the United States, we have millions of people who are living

without adequate housing or with none at all.

- What are the different types of housing found in the United States? What is the history of the different types of housing? For example, when did condominiums become popular? What do you see as the trend of the future in the different types of housing? What occupations depend on housing? How do these occupations effect housing? How can planning for housing impact on other areas, such as water and air, transporation, or health?

- What are a typical family's housing costs? What are the trends? What about your family's housing? What are your dreams and plans for housing?

- How many homeless people are there in the United States? How does homelessness today compare with homelessness ten years ago? What is being done to reduce homelessness? What are possible solutions to eliminate homelessness?

WEALTH The United States is among the wealthiest nations in the world, yet many of our citizens live in poverty. The challenge of reducing or eliminating poverty is a major problem throughout the world.

- How wealthy is the United States? How does the wealth of the United States compare with the wealth of other countries? How is wealth created? What is the history of wealth, and the trends for the future? How does our wealth relate to the declining quality of air and water?

- How is the wealth in the United States distributed. Why is it distributed this way? Who are the wealthiest individuals? What are the areas of greatest wealth, and why? How could wealth be more evenly and fairly shared? How does the amount and distribution of wealth relate to our growing problems of homelessness and hunger?

- How many people in the United States live in poverty? In the world? What is considered the poverty level in the United States? What are the trends? Where is the greatest poverty and why? How is the problem of poverty being addressed? How is hunger related to poverty?

HEALTH It is important for you to feel good, and your physical and mental health is the key. Good health depends upon how we eat and how we exercise. It depends on our care in dealing with personal hygiene, disease, and infection. Our health can

◆ How is your own health? Are you physically fit? What are your personal nutritional habits? What are your family habits? What can you do to improve the health of you and your family? What are the health problems of your community? How is the quality and quantity of water and air in your home and city? Is there pollution? How much water do you use and waste? Are there shortages?

◆ How healthy are United States citizens? What are the major health benefits and threats in our country? What are the trends, and how could these be improved? Where in the United States are there water shortages? How can we reduce demand or conserve water resources? What is "acid rain" and its effects? How can transportation, the way people move around, affect our health?

◆ How is the state of worldwide health? What is the status of disease, such as AIDS, throughout the world? What are the trends, and what can be done to avert health disasters? How has the atmosphere of Earth changed? What is the extent of air and water pollution in the world, and why? How can we improve the air and water quality for Earth?

POPULATION The population of Earth is very large and growing. The number of people in different parts of the world varies. The rates of change in the number of people also varies for different countries. Many other problems of the world, such as food, water, air quality, and transportation are affected by the size of the population.

◆ What is the population of Earth? How has the size of the population changed historically? What is the trend? How does its size affect our water and air? Do we have enough food to feed the population? Are our transportation systems adequate?

◆ How is the population of the world distributed? Which countries or cities have

impact upon our work and recreation. Though modern science and medicine have improved our wellness, disease is still a major problem in the United States and the world. Our health can be affected by the supply of clean water and air.

the greatest number of people? Where is population density the greatest or least? What is the population of your state or city? How is the production of food distributed, and how does this relate to the population distribution? How does the problem of hunger relate to population densities?

◆ How are the population data changing for different parts of the world? Where is population growth a major problem, and why? What is the future population of Earth likely to be? How might the quality of life be affected? How can we avoid problems of over-

population? What are the trends in food production and distribution? How will this affect hunger and starvation in the world?

ENDING UNIT 1

PORTFOLIO BUILDER

Now that you have completed Unit 1, refer to page 21 of your Portfolio Builder. You will find directions to begin building your own personal portfolio from items you made in Unit 1.

SURVEYS AND DATA ANALYSIS

Potato chips, the largest segment of America's snack industry, were invented in 1853 by a Native American, Adirondack chief George Crum. He was a chef at the fashionable Moon Lake House Hotel, a resort in Saratoga Springs, New York, when one of the guests, railroad magnate Commodore Cornelius Vanderbilt, complained that his fried potatoes were "too thick" and "not salty enough."

Chief Crum sarcastically decided to give the Commodore what he asked for. He sliced potatoes paper-thin, fried them to a crisp in boiling oil, and "Saratoga Chips" were born. What was intended as a prank has turned into a billion-dollar industry. Today the average American enjoys about six pounds of potato chips per year.

ABOUT THIS UNIT

In this unit, you will see how mathematics can be used to organize and interpret data gathered from a survey. Surveys are widely used to gather information concerning the circumstances, likes and dislikes, or plans of groups of people. In business and advertising, politics and government, work and recreation, and health and consumer affairs, surveys are often used to help plan for the future. You will be designing and using a survey to help you explore one of the four unit problems listed below.

▶ **UNIT PROBLEMS**

Four pairs of problems are posed below. Each of these involves one of the possible topics for the **Course Problem: housing, wealth, health, or population.**

1. HOUSING

a. In your group, plan and conduct a study of homelessness in your city and in the United States. Decide on the specific questions you want to research. Search library references or ask for information from local, state, or federal offices. Find and study data related to the problems of homeless people. From the information you find, develop a written or oral report to share what you have learned.

b. In your group, design a survey to conduct in your school that will investigate the opinions and attitudes of students about homeless people. Plan the questionnaire, collect and organize the data, and present your survey along with an oral report.

2. WEALTH

a. As a group project, study the distribution of income in the U.S. Think about several different ways you might want to show how income varies among different groups of people. What are possible sources of the data you need? Find the data, organize and interpret it, and report your findings.

b. In your group, design a survey to investigate the wealth of different groups of people. Consider such factors as age, gender, occupation, or educational background. Plan the questionnaire, collect and organize the data, and report your findings.

3. HEALTH

a. In your group, investigate health problems in your community. Contact public health offices for information. Collect data that will help you to develop a picture of the major problems or threats to health and the quality of health care. Write a report to be presented to the class.

b. As a group project, design a survey to investigate the physical health of students in your school. Focus on the connections of exercise and personal hygiene to health. For example, try to establish relationships between exercise and frequency of illness, or tooth cleaning habits and rates of cavities. Plan your questionnaire, collect and organize the data, and report your findings.

4. POPULATION

a. As a group project, study the problem of hunger and starvation in the world. Find data that will help you understand where and how much this problem occurs. Include U.S. data and make comparisons with other countries. Organize the data, perhaps using charts and graphs. Write a report to present to the class.

b. In your group, design a survey to conduct in your school to study the diets of students. Think about the questions to ask which will help you to understand their eating habits, including the nutritional quality of

their diet. Use the same survey to study the diets of some group of adults in your community. Analyze the data and compare the groups. What can you conclude about these two populations? Write a report to share your findings.

At the end of Unit 1, you should have selected one of these topics and been placed into a **Course Problem Group.** In your Course Problem Group, you should begin work on part a and complete it by the end of Activity 8. As part of the unit assessment in Activity 9, your group will develop the survey project that is part b of each topic above. By Activity 9, your group will have learned how to conduct a survey. To guide your planning, a checklist for completing your survey investigation is given below. Refer to this checklist as you develop your survey project.

Phase 1: Planning

Step 1 Develop a written plan for a survey your group will conduct.

a. Decide on the purpose of the survey. What is your goal?

b. Decide on the survey questions you will ask. Limit these to the number suggested by your teacher. Be clear about what kind of response you expect from each question. Be sure that you will get the information you want from the answers to your questions.

c. Develop a sampling plan. What size sample will you use? How should you select your sample? How will you contact your survey sample? When will you conduct your survey? How will you make sure you ask the questions exactly the same each time?

d. How will you organize, present, analyze, and interpret the data? What kind of graphs will you use to display the data?

Step 2 If time allows, present your plans to the class for their comments and feedback. Make notes about the suggestions made for improving your plan.

Step 3 Revise your plan based on the feedback from classmates and your teacher. Submit your final revised plan to your teacher.

Phase 2: Carrying Out the Plan

Step 4 Conduct your survey and collect your data.

Step 5 Make frequency tables of the data. Find the most common response by choosing and calculating the best measure of central tendency for your data. Then find the range of the data.

Step 6 Display the data with at least one graph.

Phase 3: Analyzing and Reporting the Data

Step 7 Analyze the data. Summarize your findings. Make predictions for a larger population. Write a draft of your report.

Step 8 Evaluate whether your survey provided the information necessary to meet your goal. Were your survey questions effective? Was the sample appropriate? Did the data seem to represent the population?

Step 9 Think about your approach to the investigation. How could your investigation be changed to improve the quality of the results?

Step 10 Complete a final draft of your written report. Be sure to include each of the following.

- a discussion of your purpose or goal
- a copy of your survey
- a description of your sampling plan and how it was carried out
- the data organized and summarized in frequency tables
- graphs to represent the data
- calculations to show range and central tendency
- results and conclusions
- comments about weaknesses and limitations, and ideas for improvement

EXPLORING SURVEYS

To prepare for this unit, collect as many different kinds of **surveys**, graphs, and charts as you can from newspapers, magazines, and other resources. You may either bring in the original or a photocopy. You may also want to bring the article that explains the graph or data, if one is available. You will use these throughout the unit. *You will also need to save one of these to use in your portfolio at the end of the unit.* Be ready to share with your teacher and classmates what you have found.

In this activity, you will explain how an effective survey might be designed.

▶ GROUP PROJECT A

Determine the basic elements of a survey and what makes an effective survey.

1. Study the questions in Surveys I through V that are presented below. Record and discuss your responses for each question. Compile the data for your group, and have one member report the data to your teacher so that the data for the entire class can be collected.

Survey I

◆ Did you eat breakfast this morning?

◆ Do you remember what you dreamed last night?

Survey II

Which of the following number of hours did you sleep last night?

◆ Did you sleep less than four hours last night?

◆ Did you sleep from four to six hourslast night?

◆ Did you sleep from six to eight hours last night?

◆ Did you sleep from eight to ten hours last night?

◆ Did you sleep more than ten hours last night?

Survey III

◆ Are you less than 20 years old?

◆ Are you are more than 100 years old?

◆ Would you pay a genius $5 to do your math work for you?

◆ Do you like really bad movies?

Survey IV

◆ (Ask only three people.) Who is your favorite musician?

◆ (Ask only two people.) What is your favorite movie of all time?

Survey V

◆ (Ask everyone.) What is your favorite vegetable?

2. In your group, compare surveys I through V.

   a. First, compare the survey questions. How are they different? How are they alike? As a group decide which survey asks better questions and list two reasons why you think they are better.

   b. Look at the responses, or **data**, from these surveys. How are the data different? Which are easier to understand? How could you improve the organization of the data?

   c. Think about the methods used to gather the data. Were they different? How could you improve on these methods? How could the **sample**, the group of people surveyed, be changed? How would this affect the results?

   d. Study the survey results. What are two conclusions you might reach from each survey?

   e. Look at each survey. Rewrite the questions so that the responses will give more interesting or detailed information.

3. What are some features of an effective survey? What might make a survey less effective?

4. Why conduct a survey? What can you learn from it?

5. After you have discussed the questions from Exercises 2 through 4 in your group, share your responses with the class. Be clear about each of the following.

   ◆ why surveys are conducted
   ◆ the basic parts of a survey
   ◆ the features that make a survey effective
   ◆ the sample of people that was surveyed

▶ HOMEWORK PROJECT

6. Refer to the survey questionnaire on page 22 of your Portfolio Builder.

   a. What seems to be the purpose or goal of the survey?

   b. What kinds of questions were asked? What kinds of responses will the questions produce?

   c. For what group of people was the survey questionnaire written?

   d. Conduct the survey.

   e. How many people did you survey? The more that you choose your people at random so that there is no over representation of one part of the population, the more fair or **unbiased** your sample will be. Did the type of people surveyed create a **biased** or unbiased response?

   f. Was your sample of people random? What do you think is meant by **random sample?**

g. **PORTFOLIO BUILDER** On page 23 of your Portfolio Builder, write an explanation of your survey conclusions. Include a discussion of how your data might be organized and how the results might be used.

h. Read the article titled "They're good for you, but many still shun fruit, veggies" in your Reference Section. Are your survey results similar to the results in this article? Why or why not?

►GROUP PROJECT B

Study examples of surveys.

7. Study the report entitled, "A Smoker at Home Harms Kids" in your Reference Section.

   a. List three things you learned about the actual survey from this article. Can you tell what questions were asked in the survey?

   b. What group of people was sampled? From what larger group was the sample chosen?

   c. What do you know from the article about the methods for gathering the data?

   d. From the article, state one conclusion of the survey. Do the data support the conclusion? What cautions when interpreting the data are suggested? Can you think of others?

8. Use the surveys and graphs that you have collected to discuss the types of questions that might have been used to gather the data. Then discuss the size and type of sample that the data was collected from, and the conclusions that the survey or graph suggests.

9. Suppose your group were to select one of the following ideas to investigate in a survey of students at your school.

   ◆ your "dream" car
   ◆ favorite school subject
   ◆ favorite flavor of ice cream
   ◆ things that irritate you
   ◆ favorite sport
   ◆ what you would change about your school
   ◆ favorite brand of athletic shoe

   a. Give an example of a good question for your survey. Why is it a good question? List three qualities of a good survey question.

   b. Give an example of a bad question. What makes it a bad question? List three qualities of a bad survey question.

   c. Which of the following samples would be the best for your survey? Why?

      ◆ all students who own cars
      ◆ all female students in geometry classes
      ◆ all students who walk past your math classroom
      ◆ all students who enter the front doors of the school during the 30 minutes before school starts

DESIGNING AND USING A QUESTIONNAIRE

In this activity, your group will plan a survey project. This survey will use a questionnaire that you will help to design. Each of you will complete the questionnaire, and the data for the class will be used later.

▶ GROUP PROJECT

Design a questionnaire that will help you get to know your classmates better.

1. Suppose the goal is to conduct a survey in order to get to know your class better.
   a. As a class, choose four or five categories of information that would help you and your classmates get to know more about each other.
   b. Once these have been chosen, in your group write one good question for each category. Try to think of

questions using each type of response listed below. Or, use one of the sample questions below.

Numerical responses - These are answered with a number. For example, "How old are you?" might be answered by, "I am 15 years old."

◆ How tall are you?

◆ What is your shoe size?

◆ How many CD's do you own?

◆ How many brothers and sisters do you have?

Interval responses - These are answered with a range of numbers. For example, "How long do you spend doing homework?" could result in the answer, "I spend between 1 and 2 hours on homework every night."

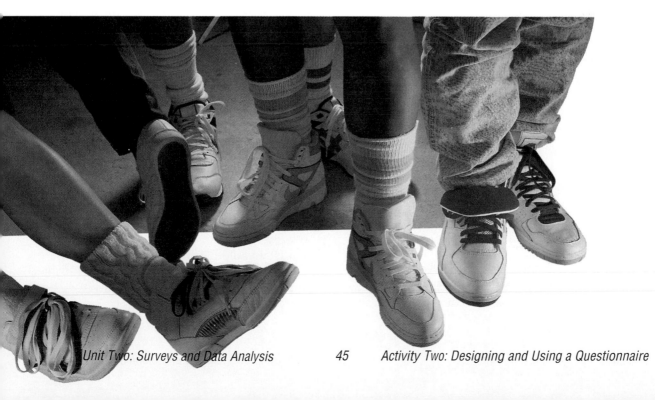

• What is the range of hours per week that you watch TV?

• What is the range of time that you go to bed?

• If you have an after-school job, what is the range of hours per week that you work?

Free responses - These are answered by telling a fact or an opinion. For example, "What is your favorite season?" could be answered with, "Autumn."

• What is your favorite _____? (For example, TV show, movie, food, sport, musical group)

• Why is _____ your favorite?

• What is the color of your eyes? Of your hair?

• In what city/state/country were you born?

2. **PORTFOLIO BUILDER** As a class, analyze and discuss the questions you wrote in Exercise 1 above. So that your questionnaire is not too long, use only two questions from each of the categories that you chose. Decide which questions are to be used. Revise the questions where necessary. Then record the categories and questions on page 24 of your Portfolio Builder. These questions will make up your class questionnaire.

3. **PORTFOLIO BUILDER** Each student should now respond to the class questionnaire. Have one group member collect and report your responses to your teacher.

Be careful, because errors made in compiling data will affect the results. This data will be used in later activities. Record the responses of your class to their questionnaires in your Portfolio Builder.

4. COMPUTER ACTIVITY Use a spreadsheet or **data base** program to record your data. Store the data so that you can use it in later activities.

5. In your group, discuss how you might organize the data in a way that tells you the most about your class. Ask yourselves the following questions.

• What could be done to summarize the responses for each question?

• How could you "picture" these summaries?

• What is the most common response for each question?

• Can you describe a "typical" student from the data?

• How do the responses vary? How much do they vary?

• What are the most unusual or rare responses?

6. From your preliminary analysis of the data, look back at your questionnaire to see how it might be improved.

• How could you change the categories to discover more?

• Are there questions that should be taken out? added?

• How could the questions be revised to make them better?

• How could you clarify or improve the purpose of the survey?

ORGANIZING AND INTERPRETING DATA

In this activity, you will use the data from your class questionnaire survey that you took in Activity 2. You will tally the data and make frequency tables. By organizing your data in a frequency table, it will make it easier to analyze and summarize it. You will also be able to use these results to predict what other samples of students might be like.

▶ **GROUP PROJECT A**

Tally data and make frequency tables.

Example 1 A survey of 25 randomly selected students was taken at a school. One of the survey questions asked the students if they were male or female. Here are the results, presented in a chart called a **frequency table**.

Gender	Tally	Frequency	Ratio	Percent
Male	𝍷𝍷 𝍷𝍷	10	$\frac{10}{25}$	40%
Female	𝍷𝍷 𝍷𝍷 𝍷𝍷	15	$\frac{15}{25}$	60%
Total		25	$\frac{25}{25}$	100%

The tally marks are recorded from the survey questionnaires. Then the tally marks are counted to find the frequency of each response. The first **ratio** compares the number of males to the total number of students who responded, 10 out of 25. This ratio can be used to find the **percent** by dividing: 10 ÷ 25 = 0.4, which is $\frac{40}{100}$ or 40%.

1. Suppose you surveyed 25 different students at this school. Would you

expect the results to be the same? Why or why not?

2. Suppose there are 1 000 students at the school where this sample was taken. How many of the students would you expect to be male? female? Why?

Example 2 On another questionnaire, students were asked to name the country in which they were born. The results are shown below.

Country	Tally	Frequency	Ratio	Percent
U.S.	𝍷𝍷 𝍷𝍷	10	$\frac{10}{40}$	25%
Mexico	𝍷𝍷 𝍷𝍷	7	$\frac{7}{40}$	17.5%
Vietnam	𝍷𝍷 𝍷𝍷 𝍷𝍷	12	$\frac{12}{40}$	30%
Korea	‖	2	$\frac{2}{40}$	5%
China		1	$\frac{1}{40}$	2.5%
Cambodia	‖‖	3	$\frac{3}{40}$	7.5%
Other	𝍷𝍷	5	$\frac{5}{40}$	12.5%

3. What percent of the students were born in the United States?
   a. This is the same as how many out of 100?
   b. This is the same as 1 out of how many?
   c. What percentage of the students were born in Korea? This is 1 out of how many?
   d. Suppose the school has a total of 2 400 students. Use the data to

estimate the number of students born in each of the countries listed in the chart. How might your estimate be effected by using the data from this sample?

▶ GROUP PROJECT B

Explore your class survey data using frequency tables.

4. ![Portfolio Builder icon] PORTFOLIO BUILDER Look at the data on page 24 of your Portfolio Builder to see what you have collected from your class questionnaire. Tally the data for each question in your class survey. On page 25 of your Portfolio Builder, make a frequency table for each response. Find the totals. Having a total helps when you check for errors.

5. When you have completed the frequency tables, share your results with another group. Check to be sure that both of your groups have obtained the same results. Correct any errors.

6. ![Portfolio Builder icon] PORTFOLIO BUILDER In your group, analyze the frequency tables. Using the information from these tables, name some characteristics that you find that describe your classmates. For example, most of the students in our class have brown hair. Record these characteristics on page 26 of your Portfolio Builder.

7. Discuss the following questions in your group.
   a. Did the survey results help you to get to know your classmates better? Why or why not?
   b. Have you found all of the reasonable conclusions from the data? What are they?
   c. What are some of the other questions that you could ask?

8. Find out how many students there are in your school.
   a. Use the survey results to predict how the total school population would respond to each question.
   b. What factors might affect the accuracy of your predictions?

9. Is getting to know your classmates too general a goal? How might you make this goal more specific? What questions would you ask if you were:
   a. on the school dance committee?
   b. planning a class trip?
   c. planning a club fund-raising activity?
   d. running for class office?

▶ HOMEWORK PROJECT

10. Think of an interesting survey question. You may wish to use something related to the Unit Problems. Think about a group for which you might want to make predictions. Choose a method of getting a sample appropriate for such predictions. Then, get the responses to your question from at least 20, but no more than 50, people. Make a frequency table. Describe your sample. Was it random? State any results or conclusions. Describe how your results might be distorted or biased. Share your findings with the class.

MEASURES OF CENTRAL TENDENCY

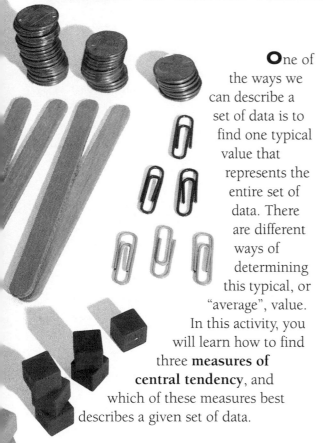

One of the ways we can describe a set of data is to find one typical value that represents the entire set of data. There are different ways of determining this typical, or "average", value. In this activity, you will learn how to find three **measures of central tendency**, and which of these measures best describes a given set of data.

▶ GROUP PROJECT A

Explore the mode, median, and mean of a set of numbers.

There are three common measures that help describe a set of data. These measures are called measures of central tendency. Your teacher will distribute some objects to you. You will use these objects to help demonstrate each of the three measures.

1. **Finding the Mode**   Sometimes the measure of central tendency is the most frequent value, called the **mode**.

   ◆ Count your objects.

   ◆ Find and join a group of students who have the same number of objects as you.

Which group has the most students? How many objects does each member of this group have? This number is the mode.

◆ If two groups have what appears to be the mode, then we say the sample has two modes, or it is **bimodal.**

◆ If three groups have the same number, then the sample is **trimodal**.

◆ If all values occur the same number of times, then there is no mode.

2. **Finding the Median**   Sometimes the measure of central tendency is the middle of the data, called the **median**. When the data are arranged in order, this is the point where exactly half of the values are greater and half of the values are less.

◆ Stand in a row in order, beginning with the person with the fewest objects and ending with the person with the most objects. Be sure that each person is in the correct place for the number of objects that they have.

◆ Next, have one person from each end of the row sit down. This will be the person with the greatest number and the person with the least number. Then, have the next person on each end sit down.

◆ Continue this pattern until you reach the middle person. The number of objects this person has is the median.

◆ If two people remain in the middle, then use the number that is halfway between the number of objects that each of them has. For example, if one person has 7 objects and the other person has 8 objects, then the median would be 7.5 or $7\frac{1}{2}$. This number can be computed by adding the two middle numbers of objects

each person has, and then dividing by two. In this case, 7 + 8 = 15.
15 ÷ 2 = 7.5.

3. **Finding the Mean**   Sometimes the measure of central tendency is where the balance point is located. If the data from two sets of objects were weights on each side of a balance, the **mean** would be the value where both sides are balanced. The word "average" is often used instead of the more mathematically precise word "mean".

- ◆ Form two lines of students. Have one line walk by the other line.

- ◆ If the person you meet has the same number of objects as you do, or one more or one less, move on.

- ◆ If the person you meet has a number of objects that is at least two more or two less than you, then share the objects so that each of you has the same number of objects, or one more or one less.

- ◆ Continue this until everyone has the same number of objects or differs only by one.

- ◆ Check. If any two students differ by more than one object, share the objects.

- ◆ If everyone has the same number of objects, you have found the mean.

- ◆ If someone does not have the same number of objects, but only differs by one, then the mean is somewhere between

the two numbers. For example, if in a class of ten students, eight have 5 objects and two have 4 objects, then the mean is between 4 and 5.

Do each of the measures of central tendency actually have to be one of the data values? Explain.

▶ **GROUP PROJECT B**

Learn to calculate and use the mode, median, and mean.

Here's how you calculate the measures of central tendency.

Mode   It is the most frequent value in the data set. In the following data set, the mode is 3, because it occurs most often.

$$7, 3, 8, 1, 4, 3, 7, 3, 12, 3, 65, 8, 9$$

Median   Arrange the values in numerical order. If there is an odd number of values, the median is the middle value. If there are an even number of values, it is the mean of the two middle values. If you rearrange our data set above, it would look like this.

$$1, 3, 3, 3, 3, 4, 7, 7, 8, 8, 9, 12, 65$$

Since there are 13 values, the median is the seventh value, or 7.

Mean   Add the values in the data set. Then divide this sum by the number of values added. For example, if you want to find the mean of the data set above, add the values. The sum is 133. Then divide by the number of values, 13. Since 133 ÷ 13 is $10\frac{3}{13}$, the mean is $10\frac{3}{13}$.

4. **PORTFOLIO BUILDER** Look again at the data gathered from your class questionnaire survey. Use the frequency tables you completed on page 25 of your Portfolio Builder.

a. Find the mean, median, and mode for each set of data. Use your calculator where appropriate. Record these with your frequency tables.

b. Which measure of central tendency would best describe the data in each frequency table? For example, if you wanted to know the most common height of the students in your class, would you use the mean, median, or mode to describe it? Explain your reasoning.

c. **PORTFOLIO BUILDER** Refer to the list of characteristics you made on page 26 of your Portfolio Builder. Now that you know the mean, median, and mode of each set of data, can you make any of your characteristics more accurate? Record your revised characteristics on page 26 of your Portfolio Builder.

d. Would these results describe the "average" student in your school? Why or why not?

e. If not, would the information describe any smaller group in the school?

f. What additional information would be necessary to describe the "average" student in your school?

g. Which measure of central tendency was used most often with the numerical data? Why?

h. Which measure of central tendency was used most often with the data that is not numerical? Why?

i. Which measure of central tendency was used most often? Why?

5. Review the newspaper article about the harmful effects of smoking in the home printed in your Reference Section. Suppose your class conducted a survey of smoking in your homes and the question "How many cigarettes are smoked per day in your home?" was included. Suppose the following data was collected for the groups in your class. You may wish to collect actual data from your class instead.

> Group 1: 0, 0, 10, 23, 36, 45
> Group 2: 0, 0, 0, 18, 23, 31
> Group 3: 0, 0, 14, 14, 24, 32
> Group 4: 4, 9, 19, 23, 27, 38
> Group 5: 0, 0, 0, 36, 49
> Group 6: 0, 8, 16, 16, 24, 32

a. **PORTFOLIO BUILDER** Find the "average" number of cigarettes smoked for each group and for all six groups combined. Find the mean, median, and mode where possible. Record this information on page 27 of your Portfolio Builder.

b.

PORTFOLIO BUILDER For each group, compare the mean, median, and mode. Which seems to be the better indicator of the center or "average?" Explain your reasoning on page 27 of your Portfolio Builder.

c. Use the six means. Find the mean of these six numbers. How does the overall mean compare with the mean of each group? Should these be the same? Why or why not?

d. **PORTFOLIO BUILDER** Refer to page 28 of your Portfolio Builder. Write about what you learned from the news article.

6. Suppose your class surveyed the color of the shirt worn by students in your school on a certain day. Here are the results.

Group 1: white, white, orange, red, blue, blue

Group 2: green, white, white, blue, pink, brown

Group 3: white, white, white, white, red, blue

Group 4: white, yellow, red, blue, orange, black

Group 5: yellow, yellow, white, white, red

Group 6: purple, yellow, orange, blue, brown, black

a. What is the most popular color in each group? What measure of central tendency would you use? Why?

b. Can you find the median? the mean? Explain why or why not.

7. Think about the survey questions listed below.

   ◆ Who is your favorite baseball player?

   ◆ When is your birthday?

   ◆ How old are you?

   ◆ How many pets does your family have?

   ◆ How old are the people living in your home?

   ◆ How long did you study last night?

   a. For each question, what kind of data will be generated?

   b. Which measure(s) of central tendency would best describe the data for each question? Why?

8. COMPUTER ACTIVITY If your computer has software for spreadsheets or a data base, find out if you can use it to calculate the mean, median, mode, or range. If so, use it to find these values for your class questionnaire data.

▶ HOMEWORK PROJECT

9. Think about what you have learned about the three measures of central tendency. When would you use each measure of central tendency (mean, median, and mode) to describe a set of data? Give examples to illustrate the significance of using one measure of central tendency rather than another. Do all sets of data have a mean? A median? A mode? How can the central tendency of non-numerical data be described?

10. Make up your own set of numerical data that has the same number for the mean, median, and mode.

WHAT'S THE PLOT?

Graphs of data can often help to make clear how the mean, median, mode, and range are related. You can review *range* on page 5 of Unit 1. If some values are changed, the effect on the mean, median, mode, and range can more easily be seen on a graph of the data. You will learn about a special kind of graph called a **line plot** in this activity. This graph will help you to tell how changes in the data can affect the measures of central tendency.

▶ PARTNER PROJECT A

Explore line plots of scores.

Statistical data can often be displayed on a number line with marks above each data type to indicate the number of pieces of data in each category. This type of display is called a line plot.

1.   **PORTFOLIO BUILDER**   Refer to page 29 of your Portfolio Builder. Study the line plots of the test scores of the three students shown. Each one shows a number line with a scale of possible test scores, from 0

through 10. The test scores are graphed using an x above the correct value on the number line. Each student has seven test scores.

a.   **PORTFOLIO BUILDER**   Find the mean, median, mode, and range of each student's scores. Use your calculator where appropriate. Mark the mean, median, and mode for Student 1 on the line plot labeled Student 1 in your Portfolio Builder. Compare these values. What would be the most fair "average" to use in determining this student's grade? Why?

b. Now mark the mean, median, and mode for Student 2 and Student 3 on the line plot, repeating the steps in Exercise 1a.

c. Notice that the sets of scores for Student 1 and Student 2 are the same except for one test. If Student 2 did not take one of the tests, is it fair to receive a grade of 0? What is the effect of this on the mean? the mode? the median?

d. Student 3s scores are identical to Student 1s scores except that Student 3 received a score of 10 on one test. What score did Student 1 get that Student 3 didn't? Student 3s mean score suggests this student deserves a grade of C. Is this fair? What about the mode and median? Discuss with your partner and be prepared to share your ideas with the class.

Activity Five: What's the Plot?

2. Suppose their teacher decides to drop the lowest score before calculating the grades.

   a. How will this change the line plots for the three students? Make new line plots showing any changes.

   b. For each student, what will be their new mode, median, mean, and range? Record these new line plots. Is this fairer? Discuss with your partner.

   c. If the highest score is dropped, how would this change the grades for each student? Is this more fair than dropping the lowest score? Explain.

3. Write a letter to another teacher explaining the effect of dropping a high or low score from a group of test scores.

4. COMPUTER ACTIVITY Find out what kind of graphs can be made using the spreadsheet or data base software on your computer. Compare the computer graphs to line plots and explain their similarities and differences.

▶ PARTNER PROJECT B

Study salary data. Be ready to share your results with the class.

5. Study the line plot of salary data for Ajax, Inc. shown below.

1992 Ajax Salaries (in thousands of dollars)

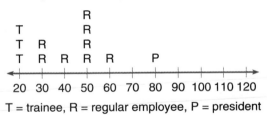

T = trainee, R = regular employee, P = president

Suppose that during 1992, the president was given a bonus of $10 000. Make a new line plot and find the range, mode, median, and mean of this new data.

6. Study the line plot of salary data for the Brown Industries Company shown .

1992 Brown Industries Salaries (in thousands of dollars)

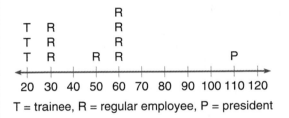

T = trainee, R = regular employee, P = president

   a. Compare the shapes of the line plots showing the 1992 salaries for Ajax and Brown Industries above. How are they alike? How are they different? How would you describe the employee salaries in each company? Why might these salary differences exist?

   b. Find the range, mode, median, and mean for the Brown salaries. Compare these with the Ajax salaries. How do they differ?

   c. Consider a $10 000 raise for all Brown employees, including the president, in 1993. How do the measures of central tendency and range values change? Record these on a new line plot.

STEM-AND-LEAF PLOTS

In Activity 5, you learned how to show the mean, median, mode, and range on a type of graph called a line plot. You can organize the same data on a kind of graph called a **stem-and-leaf plot**.

▶ GROUP PROJECT

Construct and use stem-and-leaf plots.

One way to make data more usable is to make a stem-and-leaf plot. The digit(s) in the greatest place value(s) of the data values are the *stems*. The digits in the next greatest place values are the *leaves*. For example, if all the data are two-digit numbers, the number in the tens place would be used for the stem. The number in the ones place would be used for the leaf.

1. Example 1 Refer to the World Population Data Sheet in your Reference Section. Find the Per Capita GNP (US$) for each country. This refers to the annual income per person in U.S. dollars.

A stem-and-leaf plot of the income data for the countries in Western Africa is shown below. Can you see how this graph was constructed?

PER CAPITA GNP IN WESTERN AFRICA

Stem	Leaf				
1	80				
2	40	60	70		
3	10	30	60	70	90
4	10	80			
5	00				
6					
7	10	30			
8	90				

Key: 4|80 means an income of $480.

To construct a stem-and-leaf plot, follow the steps below.

- Determine the stem and the leaf. In the example above, the stems are the hundreds digits and the leaves are the tens and ones digits of the incomes.

- Place the stems in order from least to greatest next to a vertical line. In the example above, 1| represents the numbers from 100 to 199, 2| represents the numbers from 200 to 299, and so on.

- Write the leaves on the right side of the appropriate stem.

- On your plot, write a **key** to indicate the meaning of the stem-and-leaf numbers. In this example, 4 | 80 represents an income of $480.

2. In your group, analyze the stem-and-leaf plot for the countries in Western Africa shown above. What conclusions can you reach about the incomes for the countries of Western Africa?

3. Investigate the data for each of the other four regions of Africa.
   a. Make a stem-and-leaf plot for each African region.
   b. For each plot, find the mean, median, mode, and range. Mark these values on the plot.
   c. Combine the data and make one stem-and-leaf plot for all of the countries of Africa.
   d. Find the mean, median, mode, and range and mark these on the national stem-and-leaf plot.

4. COMPUTER ACTIVITY Study your computer software for spreadsheets or a data base to see what kind of graphs can be made. Compare the computer graphs to stem-and-leaf plots of the data in Exercise 1. Discuss any similarities or differences.

   Example 2 Sometimes when data are closely grouped, using numbers on the stem that are closer together, or smaller **intervals**, can make the data easier to analyze.

Study the data for Infant Mortality Rates, the number of infant deaths per 1 000 live births, of countries in Western Africa. Use the World Population Data Sheet in your Reference Section. You will notice that most of these values are in the interval 100 to 175. A stem-and-leaf plot for this data is shown below.

INFANT MORTALITY RATES IN WESTERN AFRICA

Stem	Leaf
4	1
5	
6	
7	
8	4 6 8
9	2 9
10	
11	3 4
12	1 2 4
13	8
14	4 7 8
15	1

5. In your group, analyze the stem-and-leaf plot for Western Africa shown above.
   a. Find the mean, median, mode, and range. Mark these values on the plot.
   b. What conclusions can you reach about the death rates of infants for the countries of Western Africa?
   c. What might you predict about the rest of Africa from this data?

6. Investigate the infant mortality data for each of the other four regions of Africa.
   a. Make a stem-and-leaf plot of the infant mortality data for the other four African regions.
   b. For each plot, find the mean, median, mode, and range. Mark these values on the plot.

c. Combine the data and make one stem-and-leaf plot for all of the countries of Africa.

d. Find the mean, median, mode, and range. Mark these on the national stem-and-leaf plot.

Describe the differences. Did this method affect who won?

c. Obtain some actual scores of judges from a gymnastics or diving contest. Use line plots of the data to study the effect of dropping the highest and lowest scores.

▶ HOMEWORK PROJECT

Choose one of the following (Exercises 7 - 9) to complete on your own. Be prepared to discuss your results with the class.

7. For some sports in the Olympics, such as diving and gymnastics, the highest and lowest scores given by a judge are dropped. Study the scores given by ten judges for three gymnasts from the countries listed.

Gymnast #1 from Germany:

| 8.6 | 9.1 | 8.8 | 9.2 | 8.8 |
| 7.5 | 8.0 | 8.6 | 7.9 | 9.0 |

Gymnast #2 from Romania:

| 9.3 | 9.2 | 9.0 | 8.9 | 9.1 |
| 9.2 | 9.0 | 8.5 | 10.0 | 9.6 |

Gymnast #3 from the USA:

| 10.0 | 8.8 | 9.3 | 8.1 | 9.2 |
| 9.1 | 9.8 | 9.4 | 9.5 | 9.3 |

a. Make a line plot of the scores for each gymnast.

b. Find the mode, median, mean, and range for each set of 10 scores. Then, find the mode, median, mean, and range after dropping the highest and lowest scores for each athlete.

8. The numbers below represent the weights in pounds of fish caught by contestants in a Striped Bass Derby.

39	34	30	29	27	24	23	23
22	20	19	19	19	19	19	18
18	18	18	17	17	17	17	17
17	16	16	16	15	15	15	14
14	14	14	14	14	14	14	14
14	13	13	13	12	12	12	12
11	11	11	11	11	10	10	9
9	9	9					

a. Use the data to make a stem-and-leaf plot.

b. Obtain data from some contest in your area. Make a stem-and-leaf plot. Write a brief analysis of the data to report to the class.

9. Refer to the table labeled "Inaugurations and Deaths of U.S. Presidents," in your Reference Section.

a. Make a stem-and-leaf plot of the ages at inauguration, and another for ages at death. Which varies more?

b. Find the difference in these ages for each U.S. president. Make a stem-and-leaf plot for these data. Find the mode, median, mean, and range. How does knowing these measures help us to come to some conclusions in our analysis?

James Monroe

STATISTICAL RELATIONSHIPS FROM GRAPHS

In addition to line plots and stem-and-leaf plots, there are other ways to summarize and picture data. Today's news reports often contain graphs involving two **variables**, such as time and money. These **two-dimensional** graphs often make the overall result or **trend** easier to see, since a graph is a kind of picture of the data for both variables. In this activity, you will make and analyze various kinds of graphs, including scatter plots, bar graphs, and line graphs.

▶ PARTNER PROJECT A

Learn how to make and use scatter plots.

Scatter plots are useful for seeing if a relationship exists between two variables. To make a scatter plot, use the data given and follow the steps below.

YEAR	WORLD POPULATION (BILLIONS)
1650	0.5
1850	1
1930	2
1976	4
2016	8 (projected)

♦ Draw two **perpendicular** number lines, called axes, on a piece of grid paper. Label each axis to identify each variable as shown.

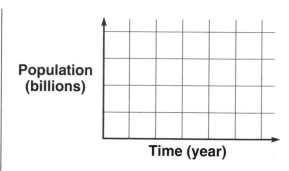

♦ Choose a **scale** for each axis that will allow the full range of values to be graphed. Mark each axis with equal intervals as shown below.

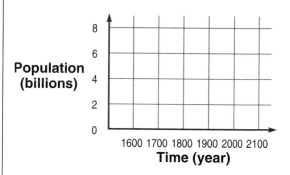

♦ Carefully graph the data. Do not connect the points.

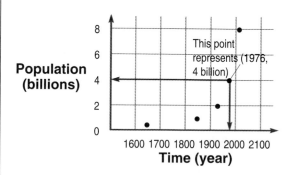

This point represents (1976, 4 billion)

Explain to your partner what the graph means to you.

1. The pairs of numbers below represent height and shoe size.

(135 cm, 6)	(155 cm, 8)
(140 cm, 7)	(150 cm, 9)
(125 cm, 7)	(180 cm, 10)
(120 cm, 8)	(190 cm, 11)

   a. Use a piece of grid paper. Label each axis, one for height and the other for shoe size. Find the range for each set of data. Then decide on appropriate scales and mark values along each axis using the ranges for each set of numbers.

   b. Plot the data points. Do you see a pattern made by the points? With your partner, write an explanation of how you think shoe size and height are related.

2. On a recent math test, students were asked to write down the average number of minutes they spent studying math each school night. Later, the teacher paired study time with the grade they earned on the test. Here are the results. Each pair represents the study time (in minutes) and letter grade for a student.

(30, A)	(5, D)	(10, D)	(15, B)	(0, F)
(15, C)	(60, A)	(0, D)	(10, C)	(20, B)
(25, A)	(25, B)	(35, B)	(45, C)	(10, A)
(30, D)	(0, A)	(30, B)	(45, B)	

   a. On grid paper, draw two perpendicular axes. Label each axis and mark intervals for each axis. Graph the data points.

   b. Look for a pattern made by the points. Is there an overall relationship between study time and grade? Justify your conclusion.

   c. Are there data that do not fit the pattern? Describe them. What do you think might be some reasons for this?

3. Think about the possible patterns that might occur with scatter plots. Some are shown below for certain variables X and Y.

i. 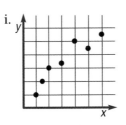   ii.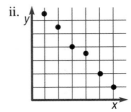

iii. 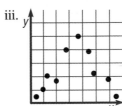   iv.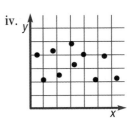

   a. For each shape, describe the relationship between X and Y.

   b. Think about some real life situations that could be related like X and Y in each example above. For example, the first scatter plot could show the relationship between the level of income and the number of years of education. Make a list of your examples to share with your class.

▶ PARTNER PROJECT B

Learn to make and use bar graphs.

Bar graphs help to make comparisons among several items in a given category. To make a bar graph, follow the steps below.

- Draw two perpendicular axes on grid paper. Label each axis to identify the variables.

- Choose a scale that will permit the full range of values to be graphed. Mark one axis with equal intervals.

- Mark the other axis with equal intervals. These intervals do not have to match those used on the other axis.

- Using the data values, carefully find the height of each bar for each item. Draw and shade each bar. Be sure to leave space between your bars.

4. Study the bar graph shown below.

a. What is being compared in this bar graph?

b. What conclusions can you make from the data shown by the bar graph?

5. Refer to your data from your class questionnaire survey on page 24 of your Portfolio Builder.

a. Use the data from one question to make a bar graph. Record your graph on page 30 of your Portfolio Builder.

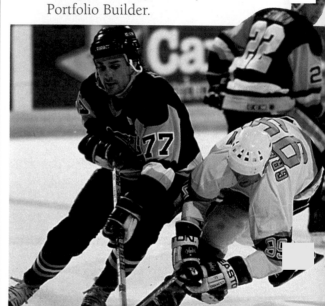

▶ PARTNER PROJECT C

Learn to make and use line graphs.

Line graphs are used to show the behavior of a variable. To make a line graph, follow the steps below.

- Draw two axes on grid paper as described before. The horizontal axis usually shows units of time, and the vertical axis represents the quantity being studied.

- Graph the points for the data. Then connect the points from left to right with line segments.

LOS ANGELES GASOLINE PRICES FOR UNLEADED REGULAR

YEAR	PRICE
1978	$ .691
1979	.943
1980	1.271
1981	1.409
1982	1.318
1983	1.215
1984	1.211
1985	1.191
1986	.956
1987	.931
1988	.910
1989	.986
1990	1.140

6. With your partner, study and discuss the following data.
   a. Make a line graph of the data.
   b. What is the year with the highest annual gasoline price?
   c. What is the greatest difference when comparing consecutive years? Why might it be different each month?
   d. How can you use the graph to find the median? the mode? the mean? the range?
   e. What is the overall pattern?

► **HOMEWORK PROJECT**

7. Refer to the "Planets of Our Solar System" chart in your Reference Section. Here are some ideas for scatter plots.
   • planet diameter and distance from the sun
   • planet diameter and number of moons
   • planet diameter and time for one rotation
   • planet diameter and time for one orbit
   • distance from the sun and number of moons
   • distance from the sun and time for one orbit
   • distance from the sun and time for one rotation
   • number of moons and time for one orbit
   • number of moons and time for one rotation
   • time for one rotation and time for one orbit
   a. Make a scatter plot using data from two different columns in the solar system chart. Are there any patterns suggested by your scatter plot?
   b. Compare and discuss your results with other students. What are the most interesting relationships found? Why?

8. The table shows the winning times (in minutes) for the men's Olympic marathon from 1896 to 1988. The times are rounded to the nearest minute.
   a. Construct an appropriate graph for this data.
   b. Find the three dates when there is no data. Why are these dates missing?

WINNING TIMES FOR THE OLYMPIC MARATHON

YEAR	TIME (MINUTES)	YEAR	TIME (MINUTES)
1896	179	1952	143
1900	180	1956	145
1904	209	1960	135
1908	175	1964	132
1912	157	1968	140
1920	153	1972	132
1924	161	1976	130
1928	153	1980	131
1932	152	1984	129
1936	149	1988	131
1948	155		

c. When did the greatest improvement in the winning time occur? Why do you think this is true?

d. What would you predict for the winning time in 1996? 2000? 2020?

9. Using newspapers or magazines, collect one example of each kind of graph you have learned about. Mount the graph on a sheet of paper. Note the type of graph and its source. Write a brief explanation of any conclusions you found from the graph.

▶ **GROUP PROJECT**

10. Select one question from the class questionnaire survey. Make a poster that shows at least three different types of graphs of your data. Write a report comparing the graphs. Explain which graph presents the information in the best way and why. You will be graded on mathematical correctness, originality, use of color, explanations, and team work.

INTERPRETING STATISTICAL GRAPHS

You have seen how several kinds of graphs can be used to organize, summarize, and show relationships about data. In this section, you will practice interpreting graphs and charts.

▶ GROUP PROJECT A

Interpret survey results.

1. A telephone survey of 600 adults was conducted in the San Francisco Bay Area in 1989. The survey questions and results were as follows.

   Question A   Do you favor or oppose legalizing all drugs, so that it is no longer a crime to use drugs such as cocaine, marijuana, and heroin?

Question B   Do you personally know anyone who has tried or who currently uses crack cocaine?

By Gender	Favor	Oppose	Don't Know
Men	23%	71%	6%
Women	15%	78%	7%

By Age	Favor	Oppose	Don't Know
18-34	17%	76%	7%
35-54	24%	68%	8%
55 & up	14%	81%	5%

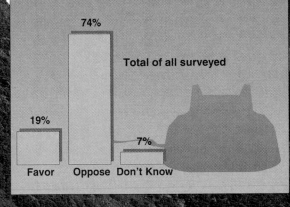

Total of all surveyed

74%

19%

7%

Favor   Oppose   Don't Know

a. What do you think was the reason for conducting this survey?

b. What do you know about the sample of people who were surveyed?

By Region	Yes	No
East Bay	32%	68%
North Bay	22%	78%
San Francisco	26%	74%
South Bay/Peninsula	18%	82%

By Age	Yes	No
18-34	32%	68%
35-54	21%	79%
55 & up	19%	81%

Total of all surveyed

76%

24%

Yes   No

c. How do you think this sample was chosen? Do you think that this sample is large enough to be a good representation of the population of the San Francisco Bay Area? Why or why not?

d. Describe the results of the survey. What do the bar graphs show? Why are the results presented by age, gender, and region?

2. How do you feel about legalizing drugs? Survey the students in your class anonymously. Summarize and interpret the data in a similar way. How do your results compare with those from the above survey?

3. This telephone survey reported a **margin of error** of ± 4 percentage points. This is sometimes misleading as it is only part of what is needed to tell us about the precision of the survey. **Confidence level** is the other part that tells us how precise the survey is.

a. Research margin of error and confidence level. Then write in your own words what you think they mean.

b. Why would the people taking the survey report a margin of error for these results?

c. Find a survey that lists a margin of error and the confidence level. Share your survey with the class.

▶ GROUP PROJECT B

Use graphs to interpret possible relationships from data.

4. The following table reports the number of car accident deaths on California freeways for each 100 million miles for cars traveling over 55 miles per hour.

a. What kind of graphs might be used to show this data? Which do you think would be the best kind of graph to use? Why?

b. **PORTFOLIO BUILDER** On page 31 of your Portfolio Builder, construct one graph to show the data. From the graph, state some patterns, trends, or relationships you see in the data.

CALIFORNIA FREEWAY FATALITIES	
YEAR	FATALITIES PER 100 MILLION MILES
1970	3.8
1971	3.2
1972	3.2
1973	3.0
1974	2.2
1975	2.2
1976	2.3
1977	2.4
1978	2.6
1979	2.5
1980	2.5
1981	2.4
1982	2.1
1983	2.0
1984	2.0
1985	1.9
1986	1.9
1987	1.8
1988	1.8
1989	1.7
1990	1.5
1991	1.3

c. **PORTFOLIO BUILDER** Find the median, mode, mean, and range for the data. Record these on page 31 of your Portfolio Builder. Then explain how knowing these might assist you in your analysis. Are any of them of no help to your analysis?

5. The results were gathered from a survey conducted by a journalism class at Exeter High School. The students asked 100 of the school's 2 000 students about their smoking habits.

EXETER HIGH SCHOOL SMOKING SURVEY	
38	Never smoked
11	Current smoker who has smoked less than 1 year
24	Current smoker who has smoked more than 1 year
18	Quit smoking less than 1 year ago
9	Quit smoking more than 1 year ago

a. In your group, discuss ways to represent the data. Think about what kind of graph is reasonable to use. Construct graphs or charts that show the appropriate patterns or relationships.

b. Many interpretations or conclusions might be made from the survey data. Discuss and write five statements based upon the data.

c. Make two predictions about the smoking habits of all 2 000 students at Exeter High. Is it reasonable to make predictions from this sample? Why or why not?

d. **PORTFOLIO BUILDER** On page 32 of your Portfolio Builder, write a brief article for the school newspaper describing the smoking survey. Include graphs or charts and other ideas from your group discussions.

IS MORE BETTER?

In this unit, you have been studying how to design a survey, collect data, organize and analyze the data, and interpret the results. Usually the data from a survey are collected from a small sample of people, but the goal is to predict something about a larger group, called the **population**. In this activity, you will explain how the size and nature of a sample group determines whether the results from a survey can be used to make reliable predictions about the larger population.

Since it can be expensive and difficult to gather data, it is important to ask how large the sample group must be. If we ask too few people, the results may not be true for the population. But on the other hand, it could be costly and time consuming to ask a very large number of people.

We might also ask how representative the sample group is. If the responses are not typical of the population, then our predictions about the population will be wrong. In this activity, you will see how a sample can be chosen more effectively.

▶ GROUP PROJECT A

Conduct a survey to explore how the choice of samples affects the results.

Suppose your class has been hired by Joan Chang, an investor who wishes to open some type of fast-food restaurant near your school. It will be your job to find out what kinds of fast food students prefer. These preferences will help her to decide which kind of restaurant to open.

Your task is to predict what type of fast food the school population might prefer. You will do this by sampling selected classes. You will use results from the survey shown below.

FAST FOOD SURVEY

Period _____ Grade _____ Age _____

Course _____

Check one: _____ male _____ female

1. What is your favorite fast food? (name one) _____

2. What is your favorite fast-food restaurant? (name one) _____

1.   First, explore within your class how the size of a sample might affect predictions about the population.

   a. Survey the members of your group. Organize the responses to show results for females, males, grade levels, and all students in your group. Based on this data from your group, make predictions about how your entire math class would respond. Think about different ways such predictions might be made with the data.

   b. Combine the data from three of the groups. Does this larger sample (consisting of three groups) result in a better prediction for the whole class? Why or why not?

   c. As a class, organize the data from all groups. How well did the data from the small samples predict the responses of the whole class? What do you think might have improved these predictions?

   d. Think about the information that you sought in the survey. Will the information from these two questions be what is needed for Ms. Chang to determine which fast food restaurant to open? Is there other information that should be collected? Suggest improvements for the questionnaire. Make all of the changes that your class agrees upon before going any further.

2.   Now do the job for Ms. Chang. This time, you will select a sample from your entire school and use the data to predict what the preferences might be for the entire student body.

   a. As a class, discuss the ideas from each of the groups. Here are some questions to ask among yourselves.

   ◆ What are the best ways to sample your school?

   ◆ What types of samples might result in data that are not representative?

   ◆ How might the size of the sample affect whether it represents the whole school?

   ◆ Should all grade levels be included?

   ◆ How can you design your sample to ensure little or no gender *bias* (prejudice towards females or males)?

   ◆ Should certain interests or habits of students be considered?

   ◆ Can you ensure that the sample is random? If not, how might that affect the results?

   b. Discuss and develop a clear sampling plan to be followed by your class. Record the steps of your plan. Try to sample students who are representative of the school population.

   c. From this plan, select the actual sample; that is, the specific students who will respond to the survey. Within the sample, are there subgroups that can be identified? Should you include additional questions in the survey to collect data about these subgroups? Will you want to keep the data for these subgroups separate? How might you use the separate data?

d. Conduct the survey. Try to implement your sampling plan as carefully as possible. Keep track of places where the actual sample may have strayed from the plan. What types of sampling errors could occur? You may want to review the sampling questions on page 40.

e. Compile the data that you collected from your school-wide sample. Organize the results in frequency tables. Construct appropriate graphs to represent the data.

f. Compare the data from your class with the data from your school-wide sample. Discuss the similarities and differences of the two groups.

g. Using the data from your school-wide sample, predict the favorite type of fast food and fast-food restaurant for the whole school. Write a one-page report for Ms. Chang. Include the following.

  ◆ a statement of the problem

  ◆ a description of the sampling plan

  ◆ discussion of why the school-wide sample was representative, large enough, unbiased, and random to give you a good estimate of what the school population truly believes

  ◆ the data displayed in a frequency table and a statistical graph

  ◆ a recommendation to Ms. Chang about which type restaurant and fast food the students will likely prefer

  ◆ comments about the possible errors in your survey study

▶ **GROUP PROJECT B**

Investigate how sample characteristics can affect results.

Suppose the goal of a survey is to see how well students can estimate the passage of 1 minute of time and to see how a distraction might affect the ability to estimate.

Separate the class into two randomly selected groups, I and II. In each group, work in pairs. Each pair will need a watch or clock to be able to measure 1 minute (60 seconds) of time. In Group I, the timer will be quiet while his or her partner sits, eyes closed, and estimates. In Group II, the timer will try to distract his or her partner during each estimation.

3. In each pair of students, one student will say "start" and time 1 minute. The second student says "stop" when she or he thinks one minute has passed. If you are the timer, don't let your partner see the clock. Record the number of seconds that actually passed.

a. Take five estimates for each student. Keep track of the order of these estimates, but don't let your partner know how he or she is doing. Record all of the results. Exchange roles and repeat the process.

b. In your pair, compare data. How do the five estimates for each of you vary? What is the range for each? Find the mean, median, and mode for each of you. How do your averages compare with your partner's? Why do you think the estimates are close or not close?

c. Combine the data for both of you, and find the mean, median, mode, and range. Make a single line plot, showing each student's results with different symbols (for example, x's and o's). How do the arrangements of the x's and o's differ?

d. Think of your pair as a sample. Use your data to predict the performance for your group (I or II). Then, compile the data for all members of this group. Find the mean, median, mode, and range. Make a line plot of the data for your entire group. How well did the results for your pair predict the actual times for the group?

e. Randomly select three pairs of students from your class and use their data to find the means, medians, and modes. How well do the data for these six students serve as a sample to predict for your class?

f. Compare the results for groups I and II. Find the mean, median, mode, and range for both. How similar are these averages? Make and compare line plots for each group. Which groupĺ seems to show the greater variation? Why?

4. Suppose that a new company called "Live!" is producing an innovative line of fashions for teens. They hire your advertising agency to develop a TV ad campaign. You want to know the following.

◆ what to put into the ads

◆ how long to make the ads

◆ in what time slots the ads should be shown

◆ who should be the target audience

◆ how the target audience will react to the ads

Your company produces four video ads to try out with people. You decide to show these ads in high traffic areas in several shopping malls around the country. Your job is to plan a survey of people who watch these videos, and to use the data to help decide which ad or ads will be used on TV.

a. Write some questions that you might ask on the questionnaire.

b. Briefly describe a sampling plan. Also, note possible biases or errors that could affect the results.

c. How might you graph the data that you collect?

d. Evaluate this survey approach. How might you change it to get better results?

▶ HOMEWORK PROJECT

5. Find a newspaper or magazine article, or photocopy a portion of a reference book that uses sampling to make a prediction. Describe what you know about the size and type of sample that was used.

CHECKING YOUR PROGRESS

In this activity, you will assess your progress in learning the mathematical ideas in this unit. First, you will complete a self-assessment for each of the topics studied.

Then your Course Problem Group for the Unit Problem will complete the study in part a and survey investigation in part b of the Unit Problems beginning on page 38.

PART I INDIVIDUAL REFLECTION

Some of the topics you have studied in this unit are listed below. Write a short paragraph describing each topic and give an example illustrating your understanding of each on pages 33 and 34 of your Portfolio Builder.

A. purposes, basic elements, and features of an effective survey

B. designing and using a questionnaire for a survey

C. designing an appropriate sampling plan for a survey

D. planning and conducting a survey

E. organizing and interpreting data in frequency tables

F. finding and using measures of central tendency

G. drawing plots and graphs

H. interpreting graphs

Use the assessment sheet on page 35 of your Portfolio Builder.

1. Indicate your understanding of the topics A – H by placing a letter for each topic on line segment 1 where appropriate.

2. Indicate how much you liked topics A – H by placing a letter for each topic on line segment 2 where appropriate.

3. Indicate how much you value what you have learned about the topics A – H by placing a letter for each topic on line segment 3 where appropriate.

PART II UNIT PROBLEM

Your group will plan and complete a study related to your Unit Problem.

4. In the Unit Overview, you were to select a topic from housing, wealth, health, and population. In your Course Problem Group, you were to plan and conduct a study of a social problem related to your topic.

   a. Meet in your Course Problem Group to complete your study. Your group should plan the questions and search for data and other information for your study. Gather and organize the data you find.

   b. Make appropriate graphs to illustrate important information about the questions you have posed.

c. Study the graphs to find patterns and form conclusions. Discuss these in your group and with your teacher.

d. Plan how your group will prepare a written report. What are the major sections of the report? Decide who will write the first drafts of each section. Set deadlines and meet to review progress. Put together your report.

Your group will complete the survey investigation related to your Unit Problem.

Much of what you will do in this activity will depend on your own planning, so it will be important for you to recall what may be needed to develop and conduct an effective survey. Refer to the checklist given in the overview for this unit.

Because your report will reflect ideas from other students, it will be important for every group member to contribute and participate positively.

ENDING UNIT 2

 Now that you have completed Unit 2, refer to page 36 of your Portfolio Builder. You will find directions for continuing building your personal portfolio from items you made in Unit 2.

RELATIVE SIZE OF NUMBERS

In 1951, Bette Nesmith was a divorced, working mother with a nine-year-old son. She also had a problem. As executive secretary to the chairman of the board of Texas Bank & Trust of Dallas, Texas, she needed to master her new electric typewriter. Not only did she make more mistakes, but she had difficulty erasing them because the ink from the carbon ribbon smeared when erased. To earn extra money, Bette also did some freelance artwork for Texas Bank. She discovered the artists never erased their mistakes, but painted over them! An idea was born. Bette used tempera waterbase paint to paint over her typing mistakes. This solved her problem.

For five years she kept her discovery a secret. After all, she was correcting her mistakes with it. Gradually other secretaries took notice, and Bette began bottling *Mistake Out*. For years she manufactured her product in her kitchen, with her son filling the bottles. Her perseverance paid off. In 1979, The Gillette Company purchased Bette's company, at the time called Liquid Paper Corporation, for $47.5 million.

ABOUT THIS UNIT

You know that a thousand is a large number, but most people do not have a clear idea of just how large it is. For example, one thousand may be a large number of ice cream cones to eat in a single day or a long line of people waiting to get in to a concert, but one thousand grains of sand might be what is emptied from a shoe after a day at the beach. Sometimes one million is not a large number. One million oxygen molecules is a very small amount of oxygen relative to the number of oxygen molecules in a coffee cup (approximately 2.7×10^{21}). The same holds true for a billion or a trillion. Can you think of examples of large numbers that are not really large in context and small numbers that are not really small in context?

Good number sense is important if you are to understand the world today. Newspaper and television reports contain very large and very small numbers. The United States federal budget is given in trillions of dollars. The world population is in billions of people. Computer speeds are in billionths of a second, and computer disk storage is in billions of bytes.

In this unit, you will see how you might think about situations that involve very large or very small numbers. You will see how to model such numbers as a way of developing a sense of the size of such quantities. You will explore the meaning of large numbers in our lives and investigate situations where understanding very large or very small numbers is important to your role as an informed citizen.

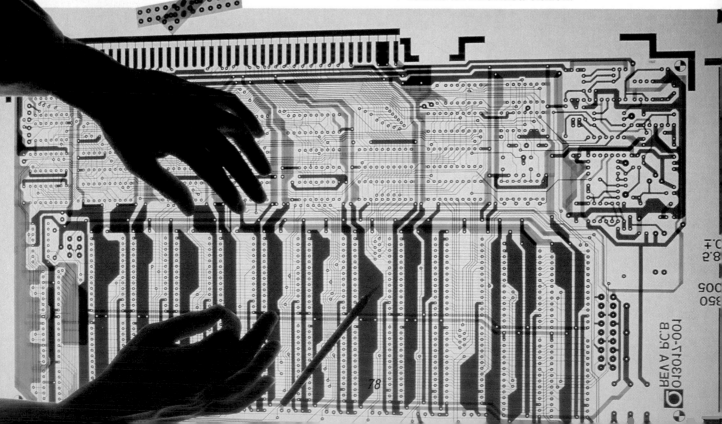

A part of good number sense is to understand the relative size of numbers. You will see that the idea of "very large" or "very small" will vary depending on the situation and how you interpret it. You will be using a form of mathematical thinking called proportional reasoning. It involves reasoning from few to many or many to few, establishing a relationship between quantities, and making judgments and predictions from known quantities to larger or smaller unknown quantities.

▶ **UNIT PROBLEMS**

Each of the topics for the Course Problem (housing, wealth, health, and population) are posed below. Early in this unit, you should select one of these topics as your Unit Problem.

1. HOUSING

The housing of our citizens is an important problem in the United States. The housing industry is one of the major elements of our economy. When you become an adult, you will need to think about your own housing.

    a. Investigate the size or extent of housing in the United States. How many housing units are there? How many are occupied by owners versus renters? How many units are vacant? If these vacant units could be used, how might these affect the problems of the homeless? How many dollars of the the U.S. economy are invested in housing? How much money is used each year to produce new housing units? What portion of the gross national product is represented by housing?

    b. What types of housing are there and what are the frequencies of each? How many are wooden structures? How much lumber is used each year to build wooden houses? How many nails are used? What other types of building materials are used? For example, how many bricks are used each year? How much cement for mortar and concrete is used?

    c. Study various distributions of people related to housing. How many homes in the United States are single-family homes? How many dwellings are more than single-family? What is the average number of persons in each home? How has the size of houses, or the number of occupants, or the number of owners versus renters changed?

d. Study the trends related to cost of U.S. housing. For example, how has the median value of a house changed in the past 50 years? How has the median rent cost changed?

2. WEALTH

The economy of a nation is a complex, but very important, matter for its citizens to understand.

a. What are the major parts of the economy, and how many dollars are involved? What is the *gross national product (GNP)*? How has the GNP changed in the past 50 years? What seems to influence the GNP?

b. Describe the federal budget of the United States. What are the major areas of expenditure in the budget? How have these categories and amounts changed in the past decade? How have these categories and amounts changed in the past 50 years?

1962-1992 FEDERAL BUDGET DEFICIT

c. What is the federal "budget deficit", and what is the current amount of this deficit? How has it changed in the past decade? How much interest is charged annually on the federal deficit? What is the daily interest charge, per citizen? Will it be possible for us to repay this debt, and how?

d. How does the size of the U.S. economy compare to the rest of the world? You might choose three other nations and study their economies for comparisons.

3. HEALTH

Clean water and air are basic needs for good health. The quality of our water and air is directly affected by the pollution from our waste. In the United States, waste products result from many different sources. Study the types and amounts of waste products in the U.S. You might choose to focus on what the average American produces. Here are some items you might choose to use.

a. Nationwide, how much trash is created each day, each week, and each year? What is the rate, per person? How do we dispose of trash? What are the trash disposal rates in your city? At the rate trash is produced in your town, how long will it be before a new disposal site must be found? How can our disposal of trash affect our health?

b. Think about all of the litter in the United States. How much litter does the average American make in one year? How much area would the litter from the United States cover, if gathered in one site? If all the litter were picked up and combined, how big of a container would be needed to hold it? If left undisposed, how long would it be before litter would cover the U.S.? What is your community doing to help the situation? What effect has recycling had on the situation?

c. How does trash and litter pollute the environment? What are the effects on the quality of our water and air? What are the trends related to such pollution?

4. POPULATION

The population of Earth is very large and growing. The number of people in different parts of the world varies. The rates of change in the number of people also varies for different countries. Many world problems are affected by the size of the population.

Study the size and distribution of the population for selected countries. Here are some items you might choose to use.

a. What is the population of Earth? What are the populations of the ten most populated countries, and what part of the total population are they? In each of the countries, how has the size of the population changed in the past 50 years? If this size increase continues, when will their population double? How can graphs be used to see the trends for each country? How is the rate for each country changing?

b. What are the population densities for your city or the town nearest you, the city with the largest population in your state, and five other major cities in the world, such as Hong Kong or Tokyo? Why do these densities vary? For possible factors that could affect population density, you may study the geographic location, land area, quality and amount of housing, climate, water and food supply, employment opportunities, transportation and traffic, birth and death rates, health conditions and medicine, waste removal, environmental pollution, poverty levels, crime rates, educational opportunities, recreation, or exports and imports. Notice that while many of these factors are affected by population size, many also affect population size.

c. How is the quality of life affected by high population density?

What problems can occur when many people live and work in a relatively small area? What advantages might there be?

Suggestions and directions for organizing and developing your project are given below. You may want to check off each step as you complete it.

Phase 1: Planning

Step 1 Develop an overall plan for solving your problem.

Step 2 Discuss your plan with your group and your teacher.

Step 3 Revise and improve your plan.

Phase 2: Carrying out the plan

Step 4 Gather information such as facts, figures, graphs, and newspaper and magazine articles.

Step 5 Organize data, compute statistics, and make graphs.

Step 6 Analyze the data.

Step 7 Make conjectures, find patterns, and make conclusions.

Phase 3: Reporting

Step 8 Write a draft of your report. Reports should include the following:

◆ a statement of the problem in which you describe the questions you want to answer

◆ a summary of facts, measures, data, and information used

◆ all computations used to prepare results

◆ any diagrams, graphs, newspaper or magazine articles, or computations needed

◆ a list of the resources you used

◆ your suggested solution to the problem

◆ a summary of your results and conclusions

Step 9 Evaluate your work, reflect on your approach, and make improvements.

Step 10 Revise and prepare a final written report.

HOW BIG IS A THOUSAND?

In order to understand the size of very large numbers, it might be better to first comprehend the size of numbers in the thousands. In this activity, you will move through stations to investigate problems that involve smaller samples of objects to make predictions about quantities of one thousand. You will also be using **proportional reasoning** to solve problems involving numbers in the thousands, tens of thousands, and hundreds of thousands. You will be making **estimates** for your answers. An estimate is a close approximation of the value, amount, size, weight … of something. An estimate is based upon a strategy and calculation rather than a random guess.

▶ GROUP PROJECT A

Explore the size of one thousand.

1. **PORTFOLIO BUILDER** You will find the supplies you will need at each station. Complete the activity at each station and record the information you gather on pages 37 and 38 in your Portfolio Builder.

Station 1 Have you ever entered a contest where you tried to guess the number of

objects in a closed and sealed container, such as beans in a jar? For this investigation, you will find a jar containing a certain number of the same object.

a. **PORTFOLIO BUILDER** Develop a strategy for estimating the number of objects in the jar without counting all of the objects. Using this strategy, make an estimate. Discuss your estimate with your group and come to a concensus. Then record your group's estimate along with your strategy and any computations you made in your Portfolio Builder.

b. Now write your group's name and estimate on the wall chart.

Station 2 Does your heart beat one thousand times in one hour? Investigate this question and write a description of the procedure you used to answer it.

Station 3 Using the sample of objects at this station, estimate how long one thousand of them would be if they were laid end-to-end. Describe your strategy.

Station 4 Estimate how much 1 000 of the objects at this station would weigh. Then estimate the weight of 10 000 and 100 000 of these objects.

Station 5 Will 1 000 basketballs one layer deep completely cover the floor of your classroom? Estimate the number of basketballs that you think would completely cover the floor in your classroom. Record your estimate and the strategy used to make it.

Station 6 Is a person who has lived 864 000 seconds very young or very old? About how many seconds have you lived?

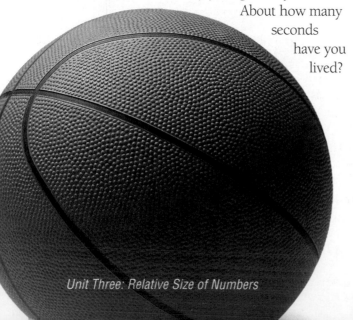

Use the materials at this station to help you with your estimate. Record your estimate. Describe how you made the estimate. Neatly write out all of the computations you used.

In developing your strategies to solve the problems in the stations, some of you may have used proportional reasoning. You may have figured out how many times your heart beats in one minute and multiplied it by the number of minutes in one hour. Or you may have figured out how many basketballs fit in one section on the floor and multiplied that number by the number of sections in the total floor.

▶ **PARTNER PROJECT A**

Explore the size of one thousand.

2. Estimate the yearly income of a person who earns $5 per hour and works eight hours a day, five days a week.

   a. Assuming the person wants to save 10% of this income, how long would it take to save enough money to pay cash for a new car costing $10 000?

   b. What **percentage** of this person's income would he or she need to save in order to accumulate $10 000 in one year? Is that reasonable? Why or why not?

   c. Many people buy a car by paying some cash and then making monthly payments on a car loan. Assume this person wants to pay $2 000 cash at the time of the purchase. At the 10% savings rate, how long will it take to save this down payment?

3. Suppose you get a new job where you are able to choose one of two payroll plans. The first one pays $1 000 per day. The second one starts at $0.01 per day, but your salary doubles each day. Explore this situation.

   a. Which plan would you choose? Why?

   b. If the job only lasts for 30 days, which payroll plan would be better? Why?

   c. Suppose the second plan doubles your daily pay every five days. How long must the job last in order to make the second plan a better deal than the first plan?

4. COMPUTER ACTIVITY Complete the following BASIC computer program to print the sequences of daily earnings and accumulated income for each of the payroll plans suggested in Exercise 3.

```
10 REM Which payroll plan?
15 LET choice 1 = 1000
20 LET choice 2 = .01
25 PRINT
   "Day  Daily 1  Total 1  Daily 2  Total 2"
30 FOR day = 1 TO 10
35 LET choice 2 = __*choice 2
40 LET total 1 = total 1 + _____
45 LET total 2 = total 2 + _____
50 PRINT
   day;choice 1;total 1;choice 2;total 2
55 NEXT day
60 END
```

a. RUN the program. Study the output to find the answer to Exercise 3a.

b. Revise the program to solve Exercise 3c.

c. Make up a different payroll offer. Revise the program to model your offer. Share the program and your results with the class.

5. Suppose a recording star earns a royalty of $0.85 for every album sold.

   a. How many albums must be sold for her to earn $1 000?

   b. If this recording star pays 68% of royalty income for taxes and other expenses, then how many albums must be sold in order for her to earn $100 000 after taxes and expenses?

▶ HOMEWORK PROJECT A

Proportional reasoning is a mathematical way of thinking about the relationship between quantities. In Unit 1, you learned to make a timeline to show the sequence of events. You might have used proportional reasoning to decide on the length of the line segment you used for one year compared to the length of the segment you used for the total number of years covered in your timeline.

6. Explain how proportional reasoning could have been used to make the estimates at each of the stations.

7. In Unit 2, you learned how data from a sample could be used to make predictions for a population. Review your results from Activity 9 of Unit 2

where you helped to determine what type of fast food might be served in a new local restaurant. Explain how you could use proportional reasoning to predict, from the sample data, how many students in your school would choose each type of fast food.

8. Describe situations when 1 000, 10 000, or 100 000 is a relatively small or large number.

▶ GROUP PROJECT B

Represent or model very large numbers.

9. In your group, make a list of at least eight things that are measured in the thousands, tens of thousands, and hundreds of thousands. Here are some examples.

   • the number of spectators at a basketball game

   • the salary of an executive in a large company

   • the altitude of a jet airplane

10. In the altitude example in Exercise 9 above, you could use a line segment one centimeter long to stand for a length of 2 000 feet. Such a segment would be a **scale drawing** of the actual item. A scale drawing is used to represent something that is too large or too small to be conveniently drawn

actual size. The ratio 2 000 ft/1 cm would be the scale. For example, if the airplane was flying at an altitude of 20 000 feet, you would need a line segment 10 centimeters long to model the altitude. Select one of the items your group listed in Exercise 9 above. Make a scale drawing of the item that shows the measures clearly. Be sure to label the scale your group is using in your drawing. Explain your drawing to the other members of your class.

▶ HOMEWORK PROJECT B

11. Bring in a non-breakable, precounted container of objects. Write the number of objects that are in your container on a piece of paper and tape it to the inside of the lid of the container. This container will be used for ongoing estimating activities in your classroom.

▶ PARTNER PROJECT B

Estimate using proportional reasoning.

12. Choose a filled container brought in by your classmates. Use proportional reasoning to determine how many items are in that container. Was your estimate close? If not, think about how you might refine your strategy here to make future estimates more accurate.

MILLIONS AND BILLIONS AND ...

In today's world, it is becoming more important for us to be able to perceive the magnitude of very large numbers. If you look around you, you will see newspaper and television news reports using large numbers. Have you ever read or heard reports about space travel in millions and trillions of miles, populations in millions and billions of people, retail sales in the millions and billions of dollars, or environmental waste being recycled in the millions and billions of pounds? In this activity, you will learn to picture what a million of something looks like. You will also solve problems involving millions, billions, and even trillions.

▶ GROUP PROJECT A

Use the length of one stride to think about the length of one million strides.

1. Suppose you went on a long hike. How far could you walk if you took one million strides?
   a. In your group, discuss what you think is a "stride." How could you determine the length of one of your typical strides?
   b. Choose one group member to model a stride length. Measure and record this stride length on the wall chart.

c. **PORTFOLIO BUILDER** Use your stride length to estimate, in reasonable units, how far a hike of approximately one million strides would take you. Record this, with your calculations, on page 39 of your Portfolio Builder.

d. Explain how you could use proportional reasoning to find out how far you could walk taking one million strides.

2. Where could you hike to if you took one million strides?

a. **PORTFOLIO BUILDER** Look at the map of the United States found on page 40 of your Portfolio Builder. Notice the scale on the map. What is the scale? Why do maps have scales?

b. On the map, draw a dot at your city or town. Then draw a circle with your city or town at the center that shows all of the places that approximately one million of your strides would take you. Underline the names of some of the places on your circle.

3. **PORTFOLIO BUILDER** Find how many strides it would take you to walk approximately one million meters. Record your work on page 39 of your Portfolio Builder.

a. On your map, can you draw a line segment to scale representing one of your stride lengths? Look back at the scale on your map to help you. Can you use a scale of 1 centimeter = 1 stride? Why or why not?

b. Now determine a scale and draw a line segment that will model how many strides will equal one million meters. Explain your work.

▶**PARTNER PROJECT**

Use stride length to think about the length of one million strides.

4. Plan your own walking trip to an interesting place that would take at least 1 000 000 strides.

a. Select actual highways to follow. Determine how far you could walk along the roads with one million of your strides.

b. Time how long it takes you to walk ten strides. Do this several times, and find a mean time. Use the mean time to estimate how long it would take you to walk, without interruption, one million strides.

▶ HOMEWORK PROJECT

5. The distance around Earth at the equator is about 40 500 kilometers.

  a. How many of your strides would you take to walk around the world along the equator?

  b. Use the average time that you found for ten strides in Exercise 4b to estimate how long your walk around Earth might take. Record your findings on page 41 of your Portfolio Builder.

  c. Adam wanted to set a world record and decided to walk the distance equivalent to the distance around Earth at the equator. How many days would it take, at a rate of 8 hours each day and 7 km/hr, for him to walk this distance?

6. Imagine a walk around Jupiter, the largest planet in our solar system. The circumference of Jupiter is about 439 800 kilometers. How many strides would you need to walk around the equator of Jupiter? About how long would it take you at a speed of 7km/hr? How many times as long will it take you to walk around the equator of Jupiter, compared to the equator of Earth? Record this

information on page 41 of your Portfolio Builder.

▶ GROUP PROJECT B

Make a scale drawing.

7. Design and make a scale drawing of the distances of the planets from the sun in the solar system. Assume that the sun is at one endpoint (zero) and the planet Pluto is at the other endpoint. Estimate the location of each planet and record this estimate as well as the scale you used, on page 42 of your Portfolio Builder. In this situation, is 1 000 a large number? What about one million? Why or why not?

▶ GROUP PROJECT C

Make estimates of magnitudes of large numbers. Make sure your answers are in units that make sense. Be prepared to explain your estimates.

8. About how many names are listed in your telephone book? Record your estimate. How did you go about making this estimate? Be prepared to explain your reasoning for determining this estimate. As a class, compare estimates and decide on the most efficient method for determining the number of names. At this rate, how many names would be listed in a telephone book that is 4 inches thick?

9. Kim likes to read novels. He reads about three novels per month, or about

1 000 pages altogether. Kim claims he has read a million pages! About how old would he have to be, if he started reading novels when he was ten years old? Is one million a large number in this situation? Why or why not?

10. How old would you be if you were one million seconds old? One billion seconds old? One trillion seconds old? Is one million seconds a long time compared to one billion seconds? Explain.

▶ GROUP PROJECT D

Explore the size of one trillion.

11. To help you get an idea of how large one trillion is, read this excerpt published in the *Republic* magazine in 1986.

"To get a picture of how much $1 trillion dollars can buy...an example using the Midwest... With...$1 trillion, we could build a $75,000 house, place it on $5,000 worth of land, furnish it with $10,000 worth of furniture, put a $10,000 car in the garage - and give all this to each and every family throughout a six state... region... Kansas, Missouri, Nebraska, Oklahoma, Colorado and Iowa. ...we could still have enough left out of our trillion dollars to build a $10-million hospital and a $10-million library for each of 250 cities and towns throughout the six-state region. After having done all that, we would still have enough money left to build 50 schools at $1 million dollars each for 500 communities in this region. And after having done all that we would still have enough left out of our trillion to put aside at 10% annual interest a sum of money that would pay a salary of $25,000 per year for 10,000 nurses, the same salary for 10,000 teachers, and an annual cash allowance of $5,000 for each and every family throughout that six-state region, not just for one year but forever."

a. Is one trillion a big number?

b. For your state, what similar goods and services would one trillion dollars buy?

12. How much money could you spend on something you really want? Could you spend $1 000 000?

a. Choose one of the dream shopping sprees described below, or make up one of your own. Spend $1 000 000.

Dream 1: Your ideal sound system.

Select all of the components it would take to have the perfect sound system. Design the ideal sound room. Include 1 000 of your favorite tapes and/or CD's.

Dream 2: The perfect wardrobe.

Select the clothing you need for every occasion (school, work, formal, casual, sports, and so on) for all four seasons. Include all of the accessories. Design your own closet and storage space, as well as an ideal dressing room.

Dream 3: The entertainment tour.

Select 12 events (for example, concerts, amusement parks, tourist attractions, and sporting events) that you would like to attend anywhere in the world, one for each month of the year. Include all travel expenses, event costs, and souvenirs. Remember to decide whether you will travel alone, or whether you will take a friend or your family.

Dream 4: Your own "wheels" for all occasions.

Select the vehicles you would want for driving to school, evening entertainment, sports, and cross-country touring. Include all optional accessories, and your usage and maintenance costs for a year.

    b. Make detailed plans for your chosen dream. Gather price information from newspaper, radio or TV ads, catalogs or brochures obtained from stores, or by direct contact with salespeople. Be sure to include any

newspaper or magazine clippings of these prices in your report.

c. Collect pictures, descriptions, and price listings of the items you pick. Use illustrations, flyers, ads, and clippings to verify prices and describe your dream items.

d. Organize your work in a display that you will present to the class or in a report. Your completed project must include facts, figures, and the total cost. In this situation is one million a large number? Why or why not?

SPACE FOR A MILLION

Have you ever thought about how much space a million of something takes up? How much space would a million people take? Would that space be more or less than the space taken by a million cars? Would that space be more or less than the space taken by a million grains of sand? In this activity, you will learn that a million can seem like a small or large number, depending on the size and arrangement of the objects.

You will also learn how to calculate the area and volume needed to hold a million objects.

▶ **GROUP PROJECT A**

Explore the size of one million.

1. **PORTFOLIO BUILDER** Use a sample of objects. Would your classroom floor hold one million of these objects one layer deep?

    a. Brainstorm with your group about possible ways to solve this problem. Write your ideas on page 43 of your Portfolio Builder.

    b. Discuss with your class all of the possible ways to solve this problem. As you continue through this activity, add to and refine the method(s) written in your Portfolio Builder.

    c. Now take some time to work on your problem, using one of your methods.

    d. Think about the measurement(s) you used. How is this measurement different from the one you used in the stride length problem?

e. In your Portfolio Builder, explain and show with drawings, measurements, and mathematical computations, how you solved the problem.

2. In Unit 2, you learned how data from a sample could be used to make predictions for a population.

a. Find the population of your state. Calculate the number of football fields you would need to hold all of the citizens of your state if they were standing on the field. Decide how much space would be required for each person. Then estimate how many football fields would be needed to hold all of the state's citizens.

b. 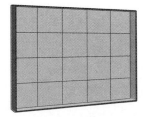 Make a scale drawing of your solution. Record your estimate and scale drawing on page 44 of your Portfolio Builder.

▶ GROUP PROJECT B

Find the area of one million objects.

Area is the number of square units needed to cover a surface. In the example below, square tiles of 1 inch are used to cover a box lid. The area of the box lid is 20 square inches.

3. Using proportional reasoning, estimate the area covered by one of your kisses. What area would one million of your kisses cover? Show all work, including sketches and measurements.

4. Choose another object from the selection your teacher provided. How much space would one million of these objects cover?

5. Look at your handprint. How many handprints would it take to cover an area of 1 000 000 square centimeters?

6. Choose one of the area problems you investigated above. Write a report about this problem. Include in your report a scale drawing, measurements used, and all mathematical calculations made.

▶ GROUP PROJECT C

Find the volume of one million objects.

7. The amount of space occupied by an object is its **volume**. To calculate the volume of the box that will hold your items, multiply the measurements of the **dimensions** you found. Volume measurements are expressed in cubic units, such as cm^3 or in^3.

a. Suppose 1 000 000 objects, like the ones you used in Group Project A, needed to be moved to another location. In order to transport them, you would need to pack these items in a reasonable sized box. Find the size box you would need to hold 1 000 000 of these items. You should be able to give the dimensions of this box.

b. Experiment with different sized boxes, and find how many of the items will fill each box.

c. Find the size box 100 of your items will fill. Be sure to list the dimensions of this box. The dimensions are the length, width, and height of the box. Make a scale drawing of the box. Find the size box that will hold 1 000 of these items. Make a scale drawing of this box. List its dimensions.

d. Use the information you have found in Exercise 7b to help you determine the dimensions of a reasonable sized box that will hold 1 million of your items. Make a scale drawing and list the dimensions.

e. Share your group's solutions with the class, explaining how you worked on the problem.

f. Calculate the volume of the boxes you found that will hold 100, 1 000, and 1 million of your items. Share your answers with your class.

8. Find the number of soda cans that you can store in a warehouse that has a floor area of 10 000 square meters and a height from floor to ceiling of 4 meters. Here are some things to consider:

a. What are some reasonable lengths and widths of a floor with 10 000 square meters?

b. Will the shape of the floor change the results? Explain.

9. What is the volume (in cubic cm) and weight of 100 drops of water? (For this investigation, you will need an eyedropper, a graduated cylinder, a balance scale, and a container of water.)

a. Use the graduated cylinder to help you find the volume of 100 drops of water. How tall would the same cylinder have to be to hold 1 000 000 drops of water? How can you use proportional reasoning to help you?

b. Use the scale provided to find the weight (in kilograms) of 1 000, 10 000, and 1 000 000 drops of water. About how much would a liter of water weigh? Look up the weight of water to check your estimates. How can you use proportional reasoning to help you?

EXPLORING POPULATIONS

Is your town or city changing? It probably is, and one of the ways it is changing may be in the number of its citizens. The world's population is growing at an alarming rate. Often this growth causes many new problems. Understanding the effects of population growth will help you and your generation deal with these problems in the future.

▶ GROUP PROJECT

Look at one of the unit problems in detail to investigate the idea of population density.

Humans are not evenly spaced on Earth. Some regions are heavily populated, while others have few people. **Population density** refers to the ratio of the number of

people per unit of land area. This is usually expressed as people per square kilometer or people per square mile. For example, the population of the United States in 1990 was 248 200 000. The area of the United States is 3 540 000 square miles. The population density is $\frac{\text{population}}{\text{area}}$. So, the population density for the United States would be 70.11299435 or about 70 people per square mile if the population were spaced evenly over the land.

1. In your group, study the population charts shown in your Reference Section.
   a. Each student in the class should choose a different state and compute the population density per square mile for five of the cities listed in that state. If your city or town is not among

the cities you worked on, find the population and land area for your city or town. Then find the population density for your city or town.

b. Each student will make a bar graph of their data. Each group will use their graphs to make comparisons among the cities. What might be some reasons for their high or low population density? What could cause a city's population density to decrease or increase?

c. Describe how a very high population density affects the type, amount, and cost of housing. Describe how it affects health care.

2. Use a map of the United States. Each student in the class will cut a piece of yarn or paper that is the same length as the bar for population density of the cities that they worked on. Label the yarn and tape one end of it to the chalkboard. Compare the lengths. Is there any pattern to where the high and low densities are located? Describe it.

3. Find the mean, median, mode, and range for population densities of the cities worked on in your group. Notice which cities are above, at, or below each measure of central tendency. Mark the measures of central tendency on your individual bar graphs. Which of the three measures of central tendency best expressed the typical population densities of the cities? Explain your choice.

▶ **HOMEWORK PROJECT**

Below are some questions about population. Research one of these and be ready to report to your group.

4. On what fraction of Earth's surface do people live? On what fraction of Earth's surface can food be grown? Are these relatively large numbers? Why or why not?

5. Suppose everyone living on Earth lived in the United States. What would be the population density?

6. Find the change in population density for your state over a 200-year period. If the region you live in has not been a state for 200 years, find the population density from the first available date. Based on this data, what would you project the population density to be in 50 years? Justify your reasoning.

7. In his classic science fiction series, *Foundation*, Isaac Asimov describes the planet Trantor, on which over 40 billion people lived! What would be the population density of Earth with this many people?

NOW IT'S EVEN BIGGER THAN A BILLION!

When we try to imagine a quantity that is in the millions or billions, it is easy to feel overwhelmed by its size. However, many situations require us to work with even larger numbers. With such numbers, it is helpful to use a way of writing numbers called **scientific notation**.

In this activity, you will learn to write large numbers using scientific notation. You will also solve problems with large numbers written in scientific notation. Most of the exercises in this activity will involve working with your calculator. You will want to try the exercises by yourself in your group. However, it will also be important for you to then share what you learn with your group members before going on to the next exercise.

▶ PARTNER PROJECT

Investigate how to name large numbers in scientific notation on your calculator.

1. The amount of fuel used by a passenger jet depends in part on the weight of the passengers and their baggage. Assume the average person and his or her baggage together weigh 200 pounds. The 20 busiest airports in the United States handle 536 million passengers per year. What is the total weight of their passengers and their baggage?

   a. [PORTFOLIO BUILDER] Turn to page 45 of your Portfolio Builder. Multiply 536 000 000 times 200 using paper and pencil. This answer is written in what is

known as **standard form**. Now find 536 000 000 times 200 on your calculator. You may be surprised to see the result. This answer is displayed in what the calculator considers scientific notation. Compare your paper and pencil answer with your calculator answer. Explain one thing that is the same and one thing that is different about the two answers.

   b. Think about the paper and pencil answer you found in Exercise 1a.

536 000 000 × 200 = 107 200 000 000 Because this number is so large, it does not fit in your calculator display in standard form. The calculator display shows a shortened version of scientific notation. A number in scientific notation is written as a product. One factor is a number that is at least 1 but less than 10. The

other factor is a power of 10. So, when the calculator displays 1.072^{11}, this actually means 1.072×10^{11} where 10 is called the **base**, and 11 is called the **exponent**.

c. We can check this answer by changing 1.072 to a fraction and then multiplying.

$$1.072 \times 10^{11} = 1.072 \times 100\ 000\ 000\ 000$$
$$= \frac{1\ 072}{1\ 000} \times 100\ 000\ 000\ 000$$
$$= 107\ 200\ 000\ 000$$

2. Write these numbers in scientific notation.
   a. 4 000
   b. 48 302 000
   c. 35 600 000 000
   d. 1 500 000
   e. 1 864 000 000 000

3. Explore how to enter numbers into your calculator in scientific notation by using the $\boxed{\text{EXP}}$ key. Look at Exercise 2a above. Since $1\ 000 = 10^3$, $4\ 000 = 4 \times 10^3$. To enter 4 000 in scientific notation on your calculator, clear the calculator first. Then enter 4 $\boxed{\text{EXP}}$ 3 to get 4. ^{03}. To check your answer, press the $\boxed{=}$ key. You should have gotten 4 000.
   a. Now use your calculator, a number from 1 to 10, and the $\boxed{\text{EXP}}$ key to enter the numbers in Exercises 2b through 2e in scientific notation.
   b. What happened when you entered these numbers?

4. What is the greatest number you can enter into your calculator in standard form? Write this number in scientific notation. Enter this number in scientific notation in your calculator to check your answer.

5. In Activity 1, you found the rate of your heartbeat. Suppose your rate is approximately the same as all the people's in the United States. There are approximately 250 million people in the United States. Find the number of heartbeats that occur in one day. Write this answer in scientific notation and in standard form.

6. Suppose the average person needs about 2 000 calories of food per day. Find the calories needed for all people in the United States per day, in standard form and in scientific notation.

7. In 1991, the average teacher salary in the U.S. was $31 304 per year. There were 3 316 428 teachers in the United States. How much money was paid in teacher's salaries in 1991? Write this in scientific notation and in standard form.

8. Approximately 42 billion gallons of water are used by people in the United States per day. What is the per person water use in the U.S. daily?

9. The U.S. national debt is about four trillion dollars. If a stack of 500 one-dollar bills is two inches high, how tall would a stack of one trillion one-dollar bills be? Explain your reasoning.

THINKING WITH VERY SMALL NUMBERS

We know that *outer* space is huge, and we need very large numbers to describe it. However, the *inner* space of microscopic objects is so tiny that the use of very small numbers is required. As with very large quantities, it is often very difficult to think about and understand these very small quantities. There might be a large number of objects, such as the cells in a drop of blood, but the area and volume they occupy can be very tiny. You can use scientific notation to help you work with small quantities as well. In this activity, you will learn to write very small numbers in scientific notation. You will also solve problems with very small numbers.

▶ **PARTNER PROJECT**

Investigate how to express small numbers in scientific notation.

1. A biologist was looking at samples of various cells under a microscope. She saw a rectangular cell. Knowing the magnifying power of the microscope, she estimated its length to be 0.0000002 cm and its width to be 0.00001 cm. What is the area of the cell?

   a. Turn to page 46 of your Portfolio Builder. Multiply 0.0000002 by 0.00001 using paper and pencil. Your answer is written in standard form. Now find 0.0000002 times 0.00001 on your calculator. 2^{-12} · 0.000000000002 Compare your paper and pencil answer with your calculator answer. The answer on your calculator shows a shortened version of scientific notation. Now write this answer in scientific notation using a power of 10.
   (2×10^{-12})

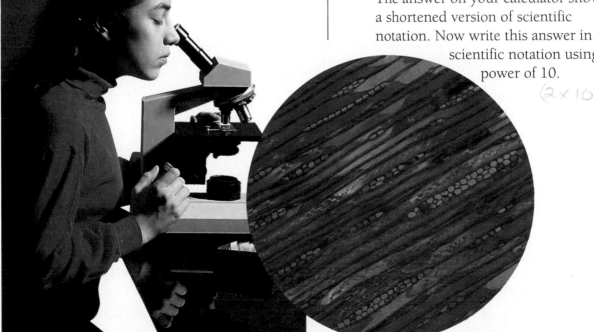

Explain one thing that is the same and one thing that is different about the three answers on page 46 of your Portfolio Builder.

b. Remember that a number in scientific notation is written as a product. Really small numbers that are between 0 and 1 are written in scientific notation as a number between 1 and 10 and thus must use negative exponents. For example,

$$0.00034 = \frac{34}{100\,000} = \frac{34}{10^5} = 3.4 \times 10^{-5}.$$

On your calculator, it would appear as 3.4^{-05}. The decimal 0.000000000002 written in scientific notation would be 2×10^{-12}. Write these numbers in scientific notation.

- ◆ 0.00342
- ◆ 0.08
- ◆ 0.000046
- ◆ 0.067890
- ◆ 0.00088772

2. Think about the area of the cell you just found. Is this cell large or small? Make a sketch of one square centimeter. Is the area of the cell more or less than one square centimeter?

3. Explore how to enter very small numbers into your calculator using scientific notation. Look at Exercise 1b above. To enter 0.00342 in scientific notation on your calculator, first clear your calculator. Then enter 3.42 [EXP] [+/-] 3 or 3.42 [EXP] 3 [+/-] to get 3.42 $^{-03}$. Then press the

[=] key to check to see if you get the number you started with.

a. Why did you enter 3.42?

b. **PORTFOLIO BUILDER** Look back to page 45 of your Portfolio Builder where you entered a very large number on your calculator in scientific notation. What is different about entering a very small number in scientific notation?

c. Use your calculator to enter the numbers in Exercise 1b in scientific notation. Check your answers by pressing the [=] key.

4. A Jelly Belly® is a gourmet jelly bean that has 4 calories and weighs 0.04 ounce. Write this weight in scientific notation. Suppose the average person could hold about 50 jelly bellies in one handful. Calculate how many calories this handful would contain. Calculate how many calories there would be if an average handful was taken by all the people in the

United States and calculate the weight, using an appropriate unit of measure, of all of these handfuls.

5. Find the thickness of a sheet of paper. Write this measurement in standard form and in scientific notation. Explain how you could use proportional reasoning to solve this problem given a ream of paper.

6. A certain computer can perform a calculation in 0.00000000025 seconds. How long would it take to perform 1 million calculations? 1 billion calculations?

▶ HOMEWORK PROJECT

7. **PORTFOLIO BUILDER** Refer to the table on page 47 of your Portfolio Builder. Look for a pattern. Think about how the decimal point is moved to change from the standard form to scientific notation. Use this pattern to complete the table.

8. The cells of our skin are constantly being replaced. Suppose the typical person loses 0.0001 gram (0.1 milligram) of skin each day. What is the mass of the skin lost by all the people on Earth (about 4.5 billion) in a day? In a year? Is this a large quantity? How does it compare to the mass of one human?

9. Find the approximate size of a snowflake. Estimate the number of snowflakes in a cubic centimeter. From this, compute how many snowflakes

might be covering one square kilometer to a depth of one decimeter. Think of some region of your state or the United States that interests you, and estimate the number of snowflakes that might fall upon it.

10. The distance between the centers of two neighboring water molecules found in ice is 2.76×10^{-10} meters. How many water molecules could you line up on a meter stick?

11. Given the number 15 000 000 000 000, go back through your work in this unit. List two examples to show when this number might be a relatively large or relatively small quantity.

12. Play this calculator game with your family. Pick any 8-digit number. Write it down, then enter it into your calculator. Using only numbers that are 10^n (for example, 100 or 10^2, $\frac{1}{10}$ or 10^{-1}, $\frac{1}{1\,000}$ or 10^{-3}, and so on), the four operation keys, and the $\boxed{=}$ key, change the number to 1. For example, change 83125679 to 1. Keep track of your steps. The person with the fewest steps is the winner. Play this game ten times and try to improve your strategy each time. After the game, share your strategies.

CHECKING YOUR PROGRESS

This activity is designed to help you and your teacher assess what you have learned in this unit. At times, you will work with another student. In Part I, you will think about what you have learned in this unit and identify one project from this unit that you think represents your best work. In Part II, you and a partner will solve some problems. In Part III, your group will complete your work on the Unit Problem.

PART I INDIVIDUAL REFLECTION

Use the assessment sheet on page 48 of your Portfolio Builder.

1. Write a paragraph describing what you think are the most important ideas you have learned in this unit and give an example to illustrate your understanding of each.

2. Identify the one item or project that you feel represents your best effort in this unit. Why did you select it?

PART II PARTNER ASSESSMENT

With your partner, select two of the problems from Exercises 3-11. Working together, develop written solutions to be turned in to your teacher. You may need to look back at your previous work in this unit or research new information to complete this section. Be sure to include all work such as sketches, tables, measurements, and mathematical computations that you used to find your solutions.

3. Would one million tennis balls fit into your classroom? How many tennis balls will your classroom hold? Is this a large number in this situation? Why or why not? Be sure you explain your strategy.

4. Identify three ways that the number one billion is used in everyday life. Select one of these ways. Develop a scale drawing to make clear just how large the number is.

5. Answer these questions about the national debt.
   - How large is the national debt?
   - How much of the federal budget is used to pay interest on the national debt?
   - Suppose each citizen of the United States had to pay an equal share of the national debt. What would be each person's share?
   - How much interest per year does each United States citizen pay?
   - What is the daily interest payment for each U.S. citizen?

6. Answer these questions about saving money during your lifetime.
   - How many seconds have you lived? Is it a relatively large number? Why or why not?
   - If you could save one cent for every minute during the next 25 years, how much money would you have?
   - At this rate, how much time would you need to save $1 000? $1 million? $1 billion? $1 trillion?

7. On the average, how much money per game is earned from ticket sales of your favorite professional football team? How much money is earned from ticket sales each season? Find the average per game

ticket sales for the National Football League. What is the total for an average regular season? At this rate, what would be the total for the next decade? Find the same amounts for another professional sport of your choice. Compare the figures for the two sports.

8. Find the total amount of money spent by the Federal government on education in grades kindergarten through 12 for a recent year. What is the average amount spent on each student? How much was it 20 years ago? 10 years ago? What would you predict would be the amount in 10 years? Find the actual amount spent on each student in your school system. How do these figures compare?

9. Find the average amount of water used by each person in your community in one year.
    a. At this rate, how much water would be needed for all of the people who live in your state for one year?
    b. What is the largest lake in your state? How many people could be supplied by the amount of water in the lake for one year? How big of a lake would be needed to supply your state for one year? The United States?

10. Suppose it takes a photographer five minutes to set up and photograph a person. How long would it take to photograph one million people one at a time if the photographer works a

normal week and year? All of the people in the United States? The world? How many people would have to be in each picture to photograph everyone in the world during a photographer's lifetime? Assume the photographer will work for 45 years.

11. Choose five occupations that you might be interested in pursuing. What are the top salaries earned by people in these occupations? Compare the top salaries of these people in different states. For example, what is the top salary for a teacher in California, Texas, New York, Iowa, and Montana? How long would it take each of the individuals with the top salary in each occupation to earn one million dollars?

PART III UNIT PROBLEM

Complete your project.

In your group, you should complete your investigation and develop your report. Refer to the checklist on page 82 to help you. Be prepared to present an oral report to your class.

ENDING UNIT 3

PORTFOLIO BUILDER

Now that you have completed Unit 3, refer to page 49 of your Portfolio Builder. You will find directions to continue building your personal portfolio from items you made in Unit 3.

SPATIAL VISUALIZATION

eets—Sheet 3.

ar. 20, 1883.

(No Model.)

No. 274,207.

J. E. MATZELIGER.
LASTING MACHINE.
Patent

FIg 1

In the years following the United States Civil War, African-American inventors were beginning to be recognized for their inventions. One such invention, patented by a self-educated black man, revolutionized the shoe industry.

Jan Earnst Matzeliger was born in 1852 in Dutch Guiana. Not satisfied with life there, he traveled to Lynn, Massachusetts, where he worked in a shoe factory, learned English at night school, and studied physics and mechanics on his own. Armed only with his talent with machines and not much money, Jan invented a *lasting machine* – a machine to connect the upper leather portion of the shoe to the inner sole.

What was the effect of this lasting machine? Not only did factories increase production from 50 pairs of shoes a day to as many as 700, but it also cut the cost of shoes in half. In New England, the income from shoe manufacturing rose by 300 percent! And what happened to Jan Matzeliger? After years of poverty and poor health, he contracted tuberculosis, a disease that was common among shoe workers. He died at the age of 37 in 1889.

ABOUT THIS UNIT

We live in space. Earth moves in space. When we look at a starry night sky, or see space shuttle astronauts working in space above Earth's surface, we see three-dimensional objects such as planets, stars and the moon.

We have learned to see three-dimensional objects and to imagine these in our mind. In mathematics, it is important to be able to visualize three-dimensional objects and to draw and interpret two-dimensional pictures of them.

In this unit, you will investigate how certain three-dimensional objects can be pictured and measured. You will learn to draw models of three-dimensional closed figures made of polygons, called **polyhedra**. Polyhedra have flat surfaces, called **face**s, that are formed by polygons and their interiors. Two faces intersect in a segment called an **edge**. Three or more edges intersect at a point called a **vertex**.

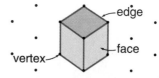

In this unit, you will be working with polyhedra, most of which are **right rectangular prisms**. In a right rectangular prism, all of the faces are rectangles. The most common right rectangular prism is one in which all of the faces are squares. This prism is called a **cube**. You will see how you might picture these in your mind and on paper. You will learn to draw a model of a cube and identify its parts, make a pattern for a cube, and draw prisms built from cubes. In your group, you will discover relationships involving volume and surface area of prisms.

▶ UNIT PROBLEMS

The problems posed below involve the four Course Problem topics of housing, wealth, health, and population.

1. HOUSING

Everyone needs shelter. Beyond the basic needs for warmth, safety, and privacy, we want to be comfortable and content. The space where we live becomes a reflection of our personalities and our psychological needs.

Do you have ideas about a "dream" residence you would like to live in? In this project, you will plan a dream residence for yourself. Your plan will feature complete scale drawings of the exterior of the building and grounds. In Unit 6, you will be able to continue this design project by making detailed interior floor plans. Your goal will be to present a complete picture of your dream residence.

a. In your group, discuss your ideas for the type of residence you would like. Help each other to identify the features to be included. Write a description of your housing needs and wants.

b. On your own, develop rough sketches of the appearance of the outside of your residence. Don't worry about drawing it accurately or to scale at this point. Focus on developing your ideas for the overall plan. Think about which three-dimensional objects will give you the maximum space. Share your sketches with your group. Give and take advice for improving the sketches.

c. Using a ruler and protractor, carefully draw detailed scale plans for the outside of your home. You may want to make either isometric drawings or orthographic projections of the exterior. You will use what you learn in this unit to help you. You may wish to include photos of exterior finishing and landscaping from magazine ads to indicate the type you would like to have.

d. How do you want the exterior walls and roof of your residence to be constructed and finished? Decide on wood siding, brick, stucco, metal siding, asphalt shingles, roof tile, or other surfaces. Contact a source of the materials, such as a lumber or brick yard. Find out prices. Compute the surface areas to be covered and estimate the costs.

e. Write and present a detailed report which will discuss your drawn plans and cost estimates. In the summary, identify the key geometric ideas you used in your project.

2. WEALTH

There are many ways spatial ideas are related to concepts of wealth. Investigate how a variety of physical objects are involved in the economy. For example, gems and minerals are used in some countries for bartering. Find unusual and interesting applications. Write a report of your ideas. Some suggestions are given below.

a. The basis for most economies is gold. Gold is stored in *ingots*. Find the size and shape of gold ingots. How many gold ingots would fill your classroom? Find the current price of an ounce of gold in the newspaper and compute the value of a classroom full of gold.

b. The gold owned by the United States government is stored as ingots in Fort Knox, Kentucky. Find the size and shape of the gold depository at Fort Knox. How much gold is stored there? How much could be stored? What is its value? How much refined gold is there in the world? If it were all processed into ingots, what size and shape depository would be needed to store it? Would it fit at Fort Knox?

c. Go to a bank and find out about the bank vault. From its size and shape, determine how many bills could be stored in the vault. Find out about the economy of your city or town. How many dollars are spent in your city or town each year? What size and shape bank vault would be needed to store all of these dollars?

d. The dome of the state capital building in the state of Georgia is covered with a layer of gold. What is the shape and surface area of the dome? How thick is the gold? What is its value? Find another public display of gold and determine its size, shape, and value.

e. Study the costs of constructing an actual office building in your city or town. Find out the size and shape of the building. What is the cost per square foot? Per cubic foot? Compare these costs with the estimated costs of building a house. Why might these costs be different?

f. Investigate the values of at least five different precious gemstones. What are the shapes into which these gems are cut? How does the size vary? What unit is used to indicate different sizes of gems? What is the relative value for different sizes for each type of gem?

3. HEALTH

Spatial shapes are involved in thinking about our health. Investigate some interesting applications of three-dimensional geometry related to health matters. Write a report of your findings. Some suggestions are given below.

a. What is the surface area of your body? Develop a technique for estimating the area. Perspiration is the body's normal method of reducing a rising temperature. Find out how the skin perspires. How much water is lost through perspiration? Use your surface area to estimate how much moisture you lose from perspiration.

b. Develop a method for estimating your lung capacity. What is the shape of a lung? How is your lung capacity related to its volume? If possible, have a discussion with a physician to determine the capacity of your lungs. How is lung capacity affected by smoking? By aerobic exercise?

c. What is the size of your heart? Find out how the surface area of your heart can be estimated from your height. What is the pumping capacity of your heart? How much blood is pumped with each contraction of the muscle? Use your pulse rate to estimate how much blood is pumped each minute, each day, each year, and during your entire life.

d. Since the heart is a muscle, it can be strengthened by regular exercise. Study how exercise can improve the capacity of the heart. Develop a personal exercise plan to strengthen your heart. Find out about heart disease in the United States. What factors contribute to heart disease? How does your own lifestyle affect the possibility that you will suffer from heart disease?

e. The human eye is a marvelous organ. What is its size and shape? What is the iris, and how does its size change? What is the size and shape of the lens of the eye? Study and make a drawing of how the lens works. Explain how eyeglasses help the eye to function.

4. POPULATION

With the population of Earth growing rapidly, we must become concerned about the amount of space available for living and for resources.

a. Investigate the amount of crop land used to feed each person in the United States. How has this changed from the 1800s to the present? How does this area compare to the areas for other countries? To the world as a whole? Based on population growth data, how much crop land might be available in the future? What might this mean?

b. Select and study an office building in your city or town. How much space is in the building? How many people work in the building? How many people visit or shop in the building each day? What might be the maximum number of people in the building at one time? What are the occupancy limits set by a fire marshall? How much space does each person have? Compare your results with another well-known building, such as the Empire State Building, the Pentagon, the Sears Tower, or with your own school or another building in your city or town.

c. One of the precious resources needed by everyone is drinking water. As water becomes scarce, one plan suggests the use of giant icebergs towed from the Arctic region to a port city where it is melted and purified. Investigate how large of an iceberg would be needed to provide water to New York City for one month. Investigate how and where this iceberg could be floated. Determine how long it would take to tow it to the port city and how much would melt while it was being towed. What size would be needed if it were towed to Miami or Los Angeles? Determine whether this would be a practical solution to a water shortage problem for the cities of New York, Los Angeles, and Miami.

BUILDING AND DRAWING THREE-DIMENSIONAL OBJECTS

Throughout your years as a student, you have learned to draw two-dimensional objects. In mathematics, it is important to be able to visualize three-dimensional objects and to draw and interpret two-dimensional pictures of them. In this activity, you will make **isometric drawings**. An isometric drawing is a drawing that shows the corner view and the top or bottom of a three-dimensional figure.

▶ PARTNER PROJECT

Explore your own spatial abilities.

1. Set out three-dimensional objects for all to see.

   a. Take turns describing the shapes of the objects. What **attributes** help to clearly describe each object?

   b. Have one partner hide the objects. Have that partner select an object secretly and describe its shape. Then have the other partner draw the object being described. Take turns drawing different objects in the same manner. Afterward, ask yourself the following questions.

   ◆ Did you make a mental picture of the object?

   ◆ Which clues were the most helpful?

   ◆ How were the objects that were described different from the actual objects?

   c. Now try it again. Have one of you select another object secretly and describe this new object to your partner. Sketch the object. Take turns guessing and sketching.

► INDIVIDUAL PROJECT

Draw cubes and prisms on isometric grids using isometric dot paper.

The arrangement of dots shown below is called an **isometric grid**. It can be used to draw a cube.

2. How is the grid made? Describe the steps you would take to make a grid of your own.

3. Look at an actual cube and the cube drawn on the grid below. Draw a cube on a grid like the one shown below.

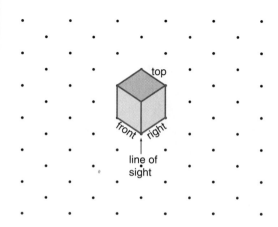

4. In an isometric drawing, one views the object from a corner. When viewing the object from the right front corner, the view is called **front-right**. The other three corners give the **front-left**, **back-left**, or **back-right** views. Notice in the picture above, the words front and right are included so that the view can be easily identified.

   a. A cube has six surfaces, called **faces**. What geometric shape is a face on a cube?

   b. How many faces do you see in the drawing of the cube? How many faces are hidden?

   c. Draw a cube on isometric dot paper so you can see the top. Shade the top of the cube to give the drawing a three-dimensional effect. Now draw a cube so you can see the bottom. Shade the bottom of the cube to give the drawing a three-dimensional effect.

5. Each corner is called a **vertex.** The plural of vertex is vertices.

   a. Look at your cube. How many vertices are there on a cube?

   b. How many vertices can be seen in the isometric drawing of the cube?

   c. How many vertices are hidden in the drawing?

   d. Draw a cube and label the vertices with capital letters. This labeling will help you discuss the different parts of a cube.

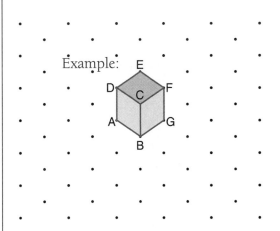

Example:

6. The line segment that is formed where two of the faces of a cube intersect is called an **edge**. For example, $\overline{CF}$ is an edge.

   a. Look at your drawing of a cube in Exercise 5d above. Name all of the edges that are shown.

   b. How many edges are showing? How many are hidden? Use dashed lines to draw the hidden edges on your drawing of the cube. Name all of the hidden edges.

   c. Now count the total number of edges on your cube.

   d. Are there different ways to label the corners to model a cube? Show as many as you can.

7. Suppose we say that the length of each edge of a cube that you drew is one unit.

   a. Use isometric dot paper to draw a larger cube whose edges are two units long.

   b. What is the length of the sum of the edges? How does this length compare with the length of the edges of the original cube?

8. Is there more than one way to represent a cube on isometric dot paper? Describe your drawing from the perspective of a person looking at your cube. How many different views can you find?

9. Make a row of three cubes. Turn your row of cubes and draw it from as many different corner views as possible.

10. 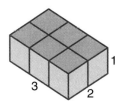 Use cubes to build each prism below.

PORTFOLIO BUILDER

   a. Build a 3-by-2-by-1 rectangular prism. These dimensions mean the following: the prism should be 3 units across, 2 units deep, and 1 unit high. Place your prism on a paper mat so that you can move it easily.

   Notice how the dimensions are included in the drawing. How many cubes are needed? Sketch this prism on page 50 of your Portfolio Builder.

   b. Turn the prism so that it is now 2-by-3-by-1. Sketch it on page 50 of your Portfolio Builder. Compare the two drawings. Explain what is the same and what is different about your drawings.

c. Turn the prism so that it is now 2-by-1-by-3. Draw it on page 51 of your Portfolio Builder. Remove the top right cube. Sketch the new polyhedron in your Portfolio Builder. Turn the paper mat so you see three other corner views. Sketch each one, labeling each drawing with the words front-right, front-left, back-right or back-left to identify which view you are representing. Share your drawings with your class.

11. Build a 2-by-2-by-2 prism. Then sketch this prism.
   a. What special kind of rectangular prism is this?
   b. Remove the two front right cubes. Sketch this new polyhedron showing at least two views.

▶ **HOMEWORK PROJECT**

12. Use isometric dot paper to draw your initials or name in cube letters.

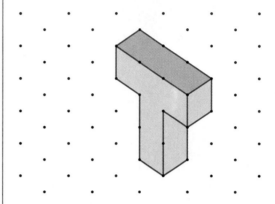

13. Make a list of all right rectangular prisms that you can find in your home. Share your list with the class.

14. Use some blocks to make a model of a familiar object such as a car or boat. Sketch your model. Share your model with your class, having them guess what you had in mind.

USING FOUNDATION DRAWINGS

In the last activity, you learned to make and use isometric drawings of three-dimensional objects. These drawings are made from a corner viewpoint. In this activity, you will explore how to interpret and draw such representations of three-dimensional objects using **foundation drawings**. These drawings show the shape of the foundation, placement, and the number of cubes that are built on this foundation.

▶ INDIVIDUAL PROJECT

Use foundation drawings to build polyhedra.

1. Refer to the foundation drawing below. The numbers in each square indicate how many small cubes are in the column to be built on that square.

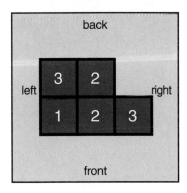

a. Use the foundation drawing to build the polyhedron with cubes. Build it on a paper mat so it will be easier to move. Label the mat with the words *front, back, left,* and *right.*

b. Make an isometric drawing of the object. Indicate the view of your drawing. Share your drawing with your group.

For Exercises 2 through 4, use the foundation drawings shown to build each polyhedron. Draw the isometric view from each of the four corners. Compare your drawings with a partner.

2. Use cubes to build this polyhedron.

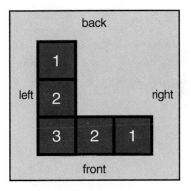

Draw and label the isometric view from each of the four corners.

3. Use cubes to build this polyhedron.

PORTFOLIO BUILDER

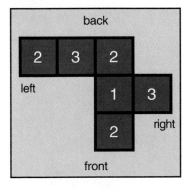

Notice that when looking at a polyhedron from certain views, not all of the cubes are always seen. On page 52 of your Portfolio Builder, draw and label the isometric view from each of the four corners.

4. Build this polyhedron.

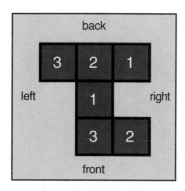

a. On page 53 of your Portfolio Builder, draw and label the isometric views.
b. Compare your results with other students.

▶ HOMEWORK PROJECT

5. Look at the isometric drawings below. Note that there are no hidden cubes. Make a foundation drawing for each.

a.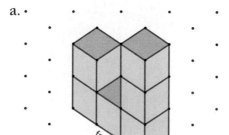

b.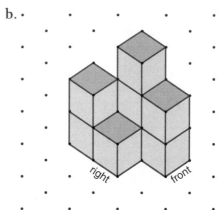

c.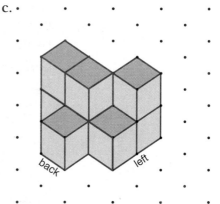

d.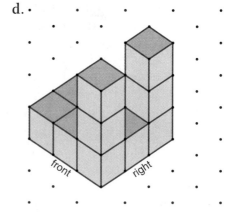

e.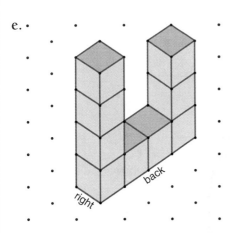

6. Look at the foundation drawing below.

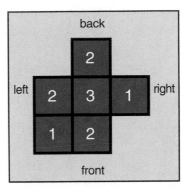

Draw and label the isometric view from each of the four corners.

GETTING ANOTHER VIEW

In the last two activities, you were making and using isometric drawings of three-dimensional objects. Designers of many different types of items often show the details of their plans with **orthographic projections**. Orthographic projections, also known as multiview drawings, enable a drafter to accurately describe on paper the size and shape of any object. This standard way to lay out a drawing makes anyone able to construct the object. In this activity, you will learn how to make this type of drawing.

Orthographic projections are drawings from the **top view**, the **front view**, and the **right side view**. These views are what you see when you look directly at one of the faces rather than viewing the object from a corner. Solid lines are drawn inside the figures to show surface edges that can be seen.

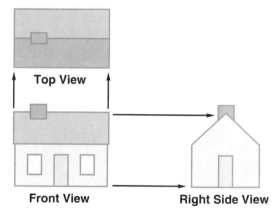

Top View

Front View   Right Side View

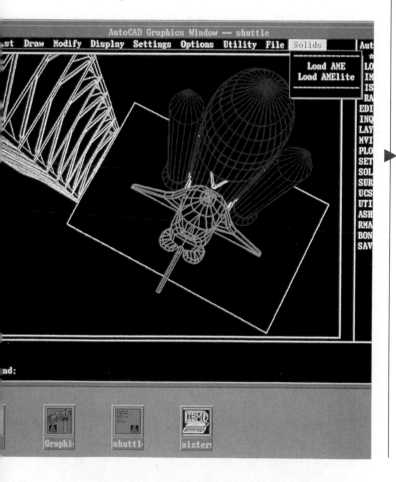

▶ GROUP PROJECT

Use orthographic projections to represent three-dimensional figures.

1. Study the foundation drawing below.

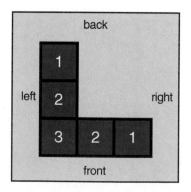

a. Make a model with cubes, using a labeled mat.

b. Study the front view below and compare it to the front view of your model.

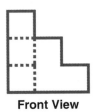

Front View

The dashed lines in this view, called **hidden lines**, are used to show surfaces that cannot be seen from the front view. Notice that looking at the front view alone will not give you enough information to recreate the object. By using an orthographic projection, more than one view is used to completely describe the solid.

c. The three views are shown below. Compare them to your model.

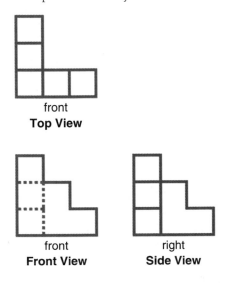

front
Top View

front
Front View

right
Side View

The front view shows the length and height of the object. The top view is projected directly above the front view. It shows the length and depth. Notice the solid lines. These are called **object lines**. Object lines show surfaces that can be seen when looking from that view. The right side view is projected directly to the right of the front view. It shows the height and depth of the object.

2. Use the foundation drawing below.

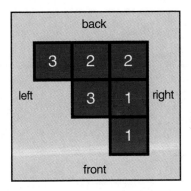

Build the model and make orthographic projections showing the top, front, and right side views. Use your model to check your drawings.

3. Refer to the foundation drawings in Exercises 3 and 4 of Activity 2.

PORTFOLIO BUILDER

a. Study the foundation drawing shown in Exercise 3. Make an orthographic projection showing all three views on page 54 of your Portfolio Builder. The isometric drawings on page 52 of your Portfolio Builder may help you with your projection.

b. Study the foundation drawing in Exercise 4. Make an orthographic projection showing all three views on page 55 of your Portfolio Builder. The isometric drawings on page 53 of your Portfolio Builder may help you with your projection.

4. Study the orthographic projections below. For each drawing, complete the following.

a. Make a foundation drawing indicating the number of blocks in each column.

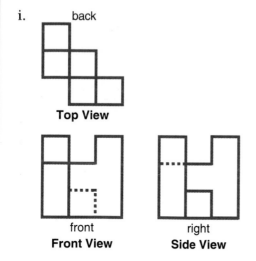

i.

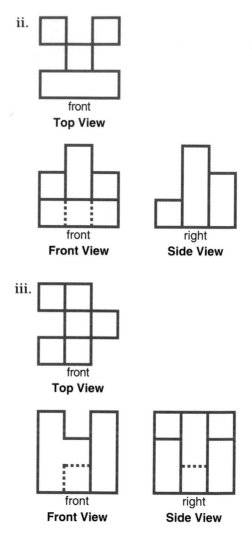

ii.

iii.

b. Then make the four isometric drawings for each. Be sure to label each view.

▶ HOMEWORK PROJECT

5. Look at the drawings below. Identify the correct orthographic view indicated by the arrow.

a.

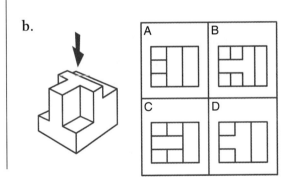

c.

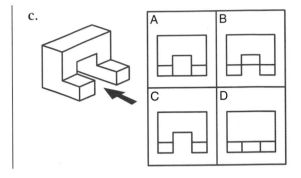

b.

CUBE PATTERNS

In this activity, you will look at patterns to solve problems about properties of cubes. For example, you will look at the dimensions of successively larger cubes. You will also review and practice using isometric drawings and orthographic projections as well as create and solve a puzzle using cubes.

▶ PARTNER PROJECT A

Use blocks to make cube patterns.

1. Build a 2-by-2-by-2 cube. Move the two front right blocks somewhere else to make a different shape. Sketch an isometric drawing of this new shape on grid paper. Turn the shape to show three other corner views and sketch each one. Then make an orthographic projection of this shape.

2. Having worked with a 1-by-1-by-1 cube and a 2-by-2-by-2 cube, find how many blocks are needed to complete the shape to make each successively larger cube. Make a chart to show your findings. What patterns do you see in this sequence of numbers?

3. Use 12 cubes. Make as many different rectangular prisms as you can using all of the cubes.

   a. 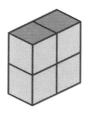 Sketch each prism, using an isometric drawing, from the front-right. Label the dimensions, checking your drawings carefully to make sure that they are all different. Put your results in the chart on page 56 of your Portfolio Builder.

   b. Compare your results with your partner. Can you make a conclusion about how many rectangular prisms are possible with 12 cubes?

   c. Use cubes to explore how many rectangular prisms are possible with 20 cubes and then 36 cubes. Discuss any conclusions you can make from the patterns you find.

▶ PARTNER PROJECT B

Make a Soma cube.

4. Two different ways of connecting four cubes are shown below.

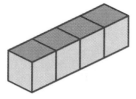

   a. There are six other ways four cubes can be connected. Build the six other figures. Make orthographic projections, foundation drawings, and isometric drawings of each group of cubes. Use a different color when drawing each piece. You will use this color code in Exercise 4c.

b. Piet Hein, a Danish engineer who created the game of Hex in the 1940s, observed that by combining the 6 shapes, that you found in Exercise 4a, and a 3-cube piece, you could form a 3-by-3-by-3 cube that he called a **Soma cube**.

c. Make the Soma cube. There are over one million ways to make the cube. Draw your solution. Use the color code from Exercise 4a to identify each Soma piece that is visible. You may need to make two or three drawings from different angles.

d. Once you have made the Soma cube, use two or three of the pieces to make the shapes below. Make

another interesting shape with two or three of the pieces. Then make an isometric drawing and an orthographic projection of it.

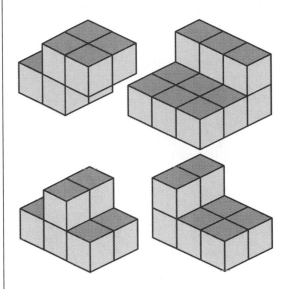

e. An interesting challenge is to "triplicate" one of the pieces. To triplicate one of the pieces, we mean that each dimension needs to be three times as long. Choose a block. Use cubes to make the same shape with dimensions three times larger. Sketch your original block and the triplicated block. Can all of the pieces be triplicated? Why or why not?

▶ HOMEWORK PROJECT

5. Invent a shape you can make with Soma blocks. Make a drawing of the shape. Share it with your class. See if your class can copy your shape using Soma blocks.

▶ GROUP PROJECT

Build cubes of different sizes.

6. In Exercise 2, you made a chart that listed the number of small cubes needed to build each successively larger cube. Build the 3-by-3-by-3 cube. Suppose you painted the cube blue on all of the outside surfaces.

   a. How many of the small cubes have:
   0 blue faces?
   1 blue face?
   2 blue faces?
   3 blue faces?
   4 blue faces?
   5 blue faces?
   6 blue faces?
   On page 57 of your Portfolio Builder, make a chart that organizes this information. Your chart should show the dimensions of each cube and the answers to each of the questions. Look at the numbers in your chart. Write in words any patterns you see.

   b. Build a 2-by-2-by-2, a 4-by-4-by-4, and a 5-by-5-by-5 cube. Answer the same questions about the faces in Exercise 6a for each and complete the chart.

7. Using your patterns in your chart, predict the answers on page 58 of your Portfolio Builder for a painted 6-by-6-by-6 cube. Use cubes to check your prediction. Extend the chart to include a painted 10-by-10-by-10 cube.

8. Write an explanation on page 58 of your Portfolio Builder that would help another student find the answer for a painted 25-by-25-by-25 cube. Include any drawings to help in your explanation.

FILLING AND COVERING

One of the most important ways that we use mathematics is to measure. Measuring is critical to almost every aspect of our lives. Think of ways that measuring is used in your life and discuss them with your group.

When working with three-dimensional objects, you may need to know how much space is inside the object or on its surface. In this activity, you will learn about **volume** and **surface area**. The volume of an object is the number of unit blocks that you would need to fill the interior space of the object. If you assume that each small block has a volume of one cubic unit, you can count the number of small blocks needed to find the volume of a polyhedron made from them. With centimeter cubes, the unit is a **cubic centimeter**.

The surface area is the number of unit squares that you would need to cover the outside of a three-dimensional object. If you assume that the square face of a cube has an area of one square unit, then you can count the number of squares that cover the surface to find the surface area. With squares one centimeter on a side, the unit is a **square centimeter**.

▶ **GROUP PROJECT A**

Create a cube net to use as a tool to find surface area.

1. Think about a cube that has been opened up. One example of the cube would look like this.

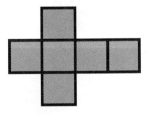

   a. How many squares are needed for the faces of the cube?
   b. The flat pattern that would fold up with no gaps or overlaps to form a cube is called a **net**. What other patterns could be drawn, cut out, and folded to make a cube?
   c. Find all of the possible cube nets. Check each net to make sure it is different from others you have made. Fold each of your nets to make sure it forms a cube.

Activity Five: Filling and Covering

d. How many different nets are possible? Make a class display of each different net. Discuss how you know that all of the possible nets have been found.

e. Find the surface area and volume of the cube that you can make from each net you have created.

2. Use cubes to build the four foundation plans below.

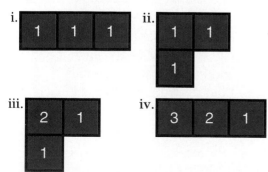

i.
| 1 | 1 | 1 |

ii.
| 1 | 1 |
| 1 |

iii.
| 2 | 1 |
| 1 |

iv.
| 3 | 2 | 1 |

a. What is the volume of each shape? Explain how you found it.

b. Draw a net of each shape. Cut out the net, fold it, then check it with the solid. Then use the net to help you find the surface area of each shape.

▶ HOMEWORK PROJECT A

3. Make a box to hold a Soma cube. Draw a net using cardboard or file folder material. Carefully cut, fold, and tape the edges to make a box. Make a Soma cube and place it in your box for storage.

▶ GROUP PROJECT B

Explore the connection between volume and surface area.

Manufacturers spend a lot of time and money designing ways to package items. Many different types and sizes of containers are used. The containers are like a net that holds these items.

Shapes having different volumes may have the same surface area. Shapes having the same volume may have different surface areas.

4. PORTFOLIO BUILDER — Refer to your previous work on page 56 of your Portfolio Builder.

a. Extend the chart to include surface area and volume.

b. What shape seems to have the least surface area? The greatest surface area?

5. Build all rectangular prisms that have a volume of 8 cubic units. What is the surface area of each shape? What shape seems to have the least surface area? The greatest surface area?

6. In your group, build several rectangular prisms using 64 cubes.
   a. Find the rectangular prism that has the greatest surface area.
   b. Find the rectangular prism that has the least surface area.
   c. Share your findings with the class.

7. Given 18 cubes, build figures that are not rectangular prisms. Find the figures with the greatest possible surface area and the least possible surface area. Find and record the surface areas and volume of the two figures. Describe how the shapes look that have the greatest possible surface area. Describe how the shapes look that have the least possible surface area. Share your results with the class.

▶ HOMEWORK PROJECT B

8. Manufacturers often package items to be displayed in an open box. Suppose you are in charge of making an open box with the greatest volume from a square piece of material. Make a pattern for the open box.
   a. Start with a piece of paper. Draw a square with sides that are 12 centimeters long. Cut out the square. Mark and cut equal square corners

from the piece. Fold the paper to make a rectangular open box.

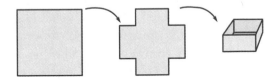

   b. What size pieces should be cut from the corners to make the box with the greatest volume? Organize an investigation. Collect data and present it in a chart. Look for patterns. What can you conclude?
   c. Investigate how to make an open box with the greatest volume for any size square piece. Record your method.
   d. How would your ideas change if you started with a rectangular piece? Investigate this situation.

9. The views below are of the same cube. Name the pairs of figures that are on opposite faces. Explain how you solved this problem.

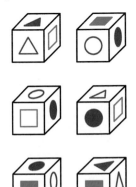

10. Use a cube net to form a cube. Mark the cube with a different letter (A, B, C, D, E, and F) on each face. In how many ways can the letter A be on the left front face? Draw on isometric dot paper each of the possible positions of the cube with A on the front-left face, showing the letters as they would appear on the front-left, front-right, and top.

11. Study the nets shown below. Draw one possible view of each cube on isometric dot paper. Show two views for each cube.

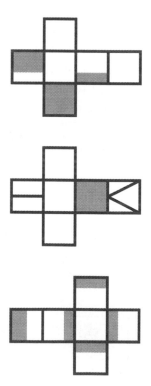

12. Use the net below. Make a sketch without folding the net that shows how the polyhedron would look. Check each other's drawings.

13. 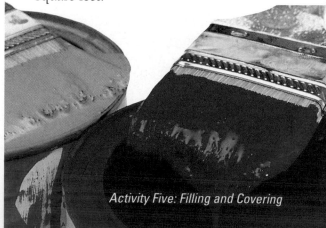 PORTFOLIO BUILDER Investigate rectangular prisms that have volumes of one million cubic units. List the dimensions of 20 of these prisms. Then find the prisms with the greatest and least surface areas. Record your work on page 59 of your Portfolio Builder.

14. How much would it cost to paint a room with four walls and a ceiling? The size of the room is 8 feet high by 10 feet long by 12 feet wide. Make a sketch of the room and label each wall. List the amount of paint that would be needed for each wall. A 2-gallon can of paint sells for $18.99 and one coat covers 25 square feet.

CHECKING YOUR PROGRESS

This activity is designed to help you and your teacher check what you have learned in this unit. At this point, you should be able to demonstrate your knowledge of spatial figures and apply ideas from this unit to use spatial reasoning to solve problems.

PART I INDIVIDUAL ASSESSMENT

In this assessment, you will use what you have learned in the unit to solve the following problem.

1. A cereal company wants to package a new cereal. The cereal containers need to be able to fit into shipping cartons and stack nicely on grocers' shelves. Each container must hold at least 36 cubic units of cereal. The container should have a minimum surface area so that it isn't too expensive to produce. Make drawings and nets of this container. Then calculate the surface area.

PART II PARTNER ASSESSMENT

In this assessment project, you will be working with a partner to construct and design a building by using what you have learned about spatial reasoning in this unit.

Your project will be graded on the following items:

◆ the level of difficulty

◆ originality
◆ mathematical correctness
◆ completeness

2. Using at least 24 cubes, build a multi-story building.
   a. Make a foundation drawing of your building.
   b. Draw the four isometric views of this building.
   c. Draw the orthographic projection of the building.
   d. Find the surface area and volume. Be sure to explain and show your work.
      e. Make a net for this building.
      f. Make a poster to display this information.
      g. Take a photograph of your poster to include in your portfolio.

PART III UNIT PROBLEM

Complete your unit problem project.

In your group, you should finish your investigation. Be prepared to present your completed work to your class.

ENDING UNIT 4

PORTFOLIO BUILDER Now that you have completed Unit 4, refer to pages 60 and 61 of your Portfolio Builder. You will find directions for continuing building your personal portfolio from items you made in this unit.

WHAT ARE MY CHANCES?

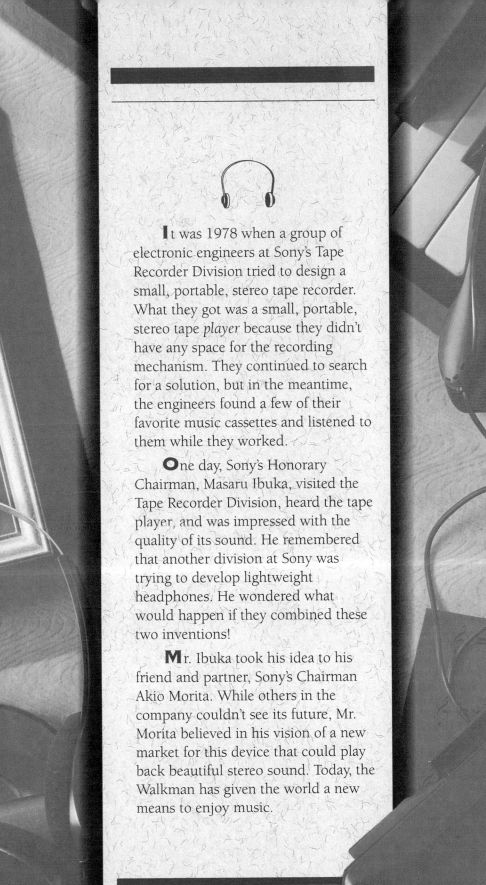

It was 1978 when a group of electronic engineers at Sony's Tape Recorder Division tried to design a small, portable, stereo tape recorder. What they got was a small, portable, stereo tape *player* because they didn't have any space for the recording mechanism. They continued to search for a solution, but in the meantime, the engineers found a few of their favorite music cassettes and listened to them while they worked.

One day, Sony's Honorary Chairman, Masaru Ibuka, visited the Tape Recorder Division, heard the tape player, and was impressed with the quality of its sound. He remembered that another division at Sony was trying to develop lightweight headphones. He wondered what would happen if they combined these two inventions!

Mr. Ibuka took his idea to his friend and partner, Sony's Chairman Akio Morita. While others in the company couldn't see its future, Mr. Morita believed in his vision of a new market for this device that could play back beautiful stereo sound. Today, the Walkman has given the world a new means to enjoy music.

ABOUT THIS UNIT

What is the probability that it will rain today? Or, that you will see an old friend? Or, that you will have exact change for a dollar? Will you have an accident, or will you win a contest? Our lives are filled with events that occur with some uncertainty. Can you think of some others?

In this unit, you will explore various situations in which probability is involved. You will see how the probability of an event can be described with a ratio. You will study various games to investigate relative frequencies and theoretical probabilities, and to see if the games are fair. You will see how the computer can be used to simulate situations using random numbers.

▶ UNIT PROBLEMS

As suggested above, most real-world events occur with uncertainty. See how chance operates within problems related to your Course Problems of housing, wealth, health, and population. From the following suggestions, you should develop your Unit Problem to involve ideas about probability.

1. HOUSING

Probability can be used to think about various aspects of risks related to housing.

a. Homeowners can experience damage to their property from a variety of randomly-occurring natural disasters. Select at least two of the following.

NATURAL DISASTERS	
earthquake	tornado
lightning	flooding
hurricane	fallen trees
hail	automobile accident
mudslide	windstorm
termites	airplane crash
forest fire	icestorm

Contact an insurance company to find out about coverage, cost, and rates of occurrence in your area for each type of disaster you selected. Try to find information, perhaps from weather records, about how often lightning strikes the ground, or hailstorms or tornadoes happen. How often do tornadoes occur in Kansas, hurricanes occur in Texas, or earthquakes occur in California? How does this affect the homeowners' insurance rates in those areas? What about hurricanes in Iowa, mudslides in New York, or icestorms in Florida? Use the information you find to decide on the chances of home damage in your area.

b. Many people buy their home using a mortgage loan. It is possible to obtain a mortgage life insurance policy. What is the purpose of this kind of insurance? What are the terms of this kind of insurance? Find out about the coverage, costs, and risks. How is probability used to describe these risks? Decide whether or not you think mortgage life insurance is a good investment. Explain your reasoning.

2. WEALTH

One of the largest parts of our economy is involved with various forms of insurance. Investigate some aspect of the economics of insurance to see how probability is used.

a. Health insurance is based on the study of probabilities. Find out how insurance companies base medical insurance plans and their costs on the analysis of probability from observed data. How do actuaries develop predictions that are used to determine the cost to the insured? How are ideas about probability used? Obtain actual data from an insurance company to demonstrate what is done. How do insurance policies that cover costs from physicians differ from those that cover costs for hospitals? What differences occur for families versus individuals? Young versus old? Rich versus poor? Why?

b. When you receive a driver's license, you will need to have auto insurance.

Investigate various coverages for auto insurance. Find out which coverages are required by law in your state. Obtain information on policies and premiums from at least three companies. Get data that shows how premiums are determined based on accident records. What are the factors that affect the size and cost of various coverages? How is probability used? Explain.

c. Life insurance is a major form of insurance in the United States. What is the size of the life insurance industry? Find information about insurance terms and premium costs for at least three different companies. Obtain data where an average life expectancy is determined. How do these vary for different groups? How is probability used? What life expectancy would you predict for yourself? What are your risks? Why?

3. HEALTH

Good health is not a certain event for anyone. Disease, injury, and even death can occur at any time. Study how the chance of health problems might be predicted.

a. One of the most serious threats to health today is AIDS. Find out what medical science knows about AIDS and its occurrence. Obtain data to demonstrate the spread of AIDS in the United States and worldwide. Based on the current incidence data and the projections for the next several years, what is the chance that someone in your age group will contract the HIV virus? Make clear how you are defining "your group." What probabilities of infection can be stated for other defined groups? What groups are the most likely to become infected, and with what chances? What factors seem to affect the incidence of AIDS? What would you predict for yourself? Why?

b. Heart disease is a serious threat to one's health. Study the incidence of heart disease in the United States. How has the incidence of heart disease changed in the past several decades? How does this data compare with the data from other countries? What is the chance that someone in your group will suffer from heart disease? Which groups are the most likely to experience heart disease, with what probabilities? Why? Which groups have the lowest chance of heart problems? Why? What are the factors that seem to affect the incidence of heart disease? What would you predict for yourself? Why?

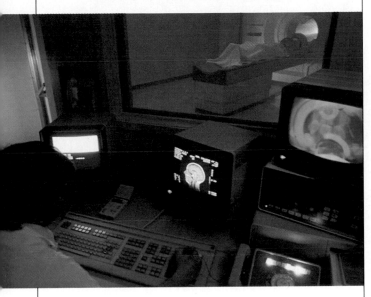

c. Cancer can be a deadly disease, yet the chance of recovery from many types of cancers is improving. Study the incidence of various types of cancers for differing groups of people. What are the chances of having cancer? What are the chances for survival? What health factors appear to influence the chances of having cancer? How has the data changed during the past few decades? What would you predict for yourself? Why?

4. POPULATION

a. The major determinant of population growth is usually the birth rate.

Study the data for birth rates in the United States each year for the past 50 years. How has the data changed? Why? What factors can affect the birth rate? How is the idea of chance or probability operating in this situation?

b. One of the factors that affect the size of any population is the death rate. Study the data for the number of deaths in the United States each year during the past 50 years. How has this number changed? Based on the size of the population, find the death rate for each year. How has this number changed? Why? What factors affect the death rate? Life expectancy for men and women in the United States has increased. Make graphs to study how the life expectancy has changed in the past 50 years. What factors affect life expectancy? How do these factors apply to you? How is the idea of probability used here?

c. Genetics is the study of how successive generations can be affected by parental traits. Identify some important human traits that can be passed from one generation to the next. Select some traits to study in your own family. Diagram how these traits have appeared in your family for as many generations as you can. Which are the dominant traits? Why? How does chance operate in the area of genetics?

FROM ZERO TO ONE

In this activity, you will think about events where chance is involved. You will learn to form a ratio to find the probability of an event. These ratios can be compared to decide which event is more likely to occur. The computer can be used to generate random numbers as a basis for chance.

▶ GROUP PROJECT A

Think about how chance occurs.

1. Have a class discussion to share ideas about chance. Think of a situation that is certain to occur. Think of a situation that will never happen.
   a. In your group, make a list of things that may occur by chance.
   b. Find articles about chance from newspapers and magazines. Discuss the articles with your group. Compare the ways that chance is involved.

FIX

c. Organize your list and articles into categories according to the chance they have of occurring.

2. **PORTFOLIO BUILDER** In your group, read and discuss each of the four problems listed below. Discuss how the idea of chance is involved in each of the problems. Talk about possible solutions for each problem. Make some reasonable guesses for each solution. You do not need to solve the problems at this time. Record your ideas on page 62 of your Portfolio Builder.

"Marriage" Problem   In ancient times, a couple had to apply to the ruler for a marriage permit. The ruler handed the couple six pieces of rope. One person had to hold the pieces of rope in the middle so that it was impossible to tell which ends belonged to the same piece of rope. Six loose ends of rope hung down on each side of the hand.

Then, on each of these sides, three pairs of loose ends were tied at random by the other person. If the result turned out to be a single, closed ring, the couple was given a marriage permit.

If not, the couple had to wait one full year to reapply. Of course, the myth was that the ring predicted the family circle and the circle of love. What is the probability of obtaining a single, closed ring?

By the ruler's law, if the couple failed to obtain a single ring for ten years in a row, they had to remain single. What is the probability that a couple would fail ten years in a row?

"Ideal Family" Problem A couple has decided that they will have four children. They feel that a family of two boys and two girls would be ideal. Assuming that they will have 4 children, what is the probability that they will have their ideal family?

"Rolling a Number Cube" Problem Suppose a game is based on rolling one number cube. Success is rolling a 1 or a 6, and failure is rolling a 2, 3, 4, or 5. What is the probability of getting exactly two successes in five rolls of the number cube?

"Baseball Batting Average" Problem Suppose that a baseball player has a batting average of .300. What is the probability

that the player will get at least three hits if the player is at bat four times in a game?

▶ HOMEWORK PROJECT A

3. Find other articles and contests from your newspapers or mail about chance. Post the articles on a bulletin board in your classroom to show different ways that chance is used. You will need to use these articles and contests later in your portfolio.

▶ GROUP PROJECT B

Explore probability ratios.

The idea of chance is made clearer when we express the ratio of the number of favorable outcomes to the number of possible outcomes. This ratio is called the probability of the event.

For example, suppose that you surveyed 20 people to find their favorite ice cream flavor. People said they liked chocolate, strawberry, vanilla, and mocha. These are the four possible outcomes. There are 8 people who report their favorite flavor as chocolate. From the ratio $\frac{8}{20}$, we could say that the probability, or chance, that a randomly chosen person in this group likes chocolate ice cream best is $\frac{2}{5}$ or 0.4. We can write this probability as:

$$P(\text{chocolate}) = \frac{8}{20}$$
$$= \frac{2}{5}$$
$$= 0.4$$

4. In your group, choose one of the following situations. Or you may want to use data you collected in Unit 2 or data of your own choosing.

- ◆ Number of radios owned by your family
- ◆ Colors of hair
- ◆ Brands of toothpaste used
- ◆ Kinds of pets
- ◆ Kinds of trees
- ◆ Types of electronic entertainment in your home
- ◆ Favorite kind of snack food
- ◆ Colors of tops worn

a. Decide on the different possible outcomes for your situation. Collect data from the other members of the class. Tally your survey data to show the results for each situation.

b. Write the probability for each outcome. What does each part of the ratio mean? From the ratios you found in your situation, which outcome is the most likely? The least likely? Make a number line. Place these ratios on the number line and label each point. You may wish to use your calculator to help you order the numbers.

c. Look at the results. Were there any events that could never happen, could happen, or could always happen? Where on a number line would these events be located?

d. Have one person from your group label and write your group's data on the chalkboard. Compare the data. Discuss some conclusions from your class results. What is the greatest possible probability? What is the least possible probability? Explain.

e. Think of events for your situation where $P(\text{event}) = 0$ and $P(\text{event}) = 1$ are true. What do these mean?

5. In the ice cream example, 8 people chose chocolate, 5 chose mocha, 2 chose strawberry, no one chose butter pecan, and the rest chose vanilla. If the event is choosing mocha, then you would write the probability of choosing mocha as $P(\text{mocha})$. Find $P(\text{mocha})$. Then find the probability for the following events:

$P(\text{strawberry})$

$P(\text{vanilla})$

$P(\text{butter pecan})$

$P(\text{chocolate or strawberry or mocha or vanilla})$

Write these probabilities on a number line. Was there any event that never happened? Was there any event that was certain to happen? What part of the number line was used?

▶ HOMEWORK PROJECT B

6. Think about everyday experiences.
   a. List at least 10 everyday experiences that can fit into the three categories of could never happen, could happen, and certain to happen.
   b. Assign a letter to each of these experiences. Make a number line. Place the letters on the number line to show the number that best describes the probability of the experiences occurring. Discuss your reasons.

7. Write each of the following words above the number line you made in Exercise 6b to match your experiences.
   A. Never
   B. Always
   C. Likely
   D. Rarely
   E. Probably
   F. Even chance
   G. Certain
   H. Often
   I. Frequently
   J. Occasionally
   K. Seldom
   L. Unlikely
   M. Less than even
   N. More than even

▶ PARTNER PROJECT

Explore relative frequencies.

8. Use a bag of colored cubes or some other item that has a mixture of colors or types in the package.
   a. What are the different colors of your items? How many of each color are in your bag? Make a chart to record the number of each color. Use your results to write a probability ratio that tells the chance of drawing each color. Record and label these results above the number line. Which color would be most likely to be selected? Least likely?
   b. Conduct a simple experiment. Without looking, randomly remove one item from the bag. Record the color. Return the item to the bag. Shake the bag. Randomly remove another item, record the color, return it to the

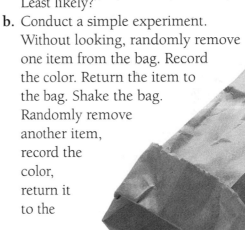

bag, and mix them up. Repeat this process 50 times. In this experiment, the fraction of the number of times each colored item is selected out of the total number of times you removed an item is called the **relative frequency.**

c. Find the relative frequency for each color of item in your bag. Record and label these results below the same number line that you used in Exercise 8a. Discuss how these ratios compare to your actual probability ratios in Exercise 8a. Have a class discussion to compare the relative frequencies from the different partner groups.

9. COMPUTER ACTIVITY With the computer, we can simulate a random process with the RND function. This BASIC program will produce 20 random numbers.

```
10 REM Random Numbers
20 FOR n = 1 to 20
30 LET x = RND(x)
40 PRINT n, x, 10*x, INT(10*x)
50 NEXT n
60 END
```

a. RUN the program several times. Study the output to see how the random number generator works. What kind of numbers result from RND? What is the range of these numbers?

b. What results from the expressions 10*x and INT(10*x)? Explore what happens in these expressions if the multiplier is changed from 10 to 6, to 2, and to 15. Later we will see that this technique will be useful to produce random numbers in a specific range.

SIMPLE PROBABILITY EXPERIMENTS

In this activity, you will conduct simple probability experiments and compare your results with the theoretical probabilities of the events. You will see how to analyze a situation to find the theoretical probability of an event. You will find ways to organize and think about all of the possible outcomes using such things as lists and tree diagrams.

▶ GROUP PROJECT A

Investigate simple probability situations.

In many situations, it is possible to state what the probability of each outcome or event should be. The list of all possible outcomes is called the **sample space**. We can use the sample space to help us determine the theoretical probability ratio. The **theoretical probability** of an event is the ratio of the number of elements in the event compared to the number of elements in the sample space.

1. In your group, discuss the following situation. A bag has four colored cubes, one each of red, green, yellow, and orange.
   a. Suppose you randomly choose a cube from the bag. What is the sample space? What is the probability of each outcome? These ratios are called the theoretical probability. What is the sum of the probabilities of all of these outcomes?
   b. Conduct an experiment using the bag of cubes above. Complete 20 trials. In each trial, draw a cube from the bag, record its color, and then return it to the bag. From the observed data, form the ratios to show the relative frequencies, or number of times a particular color of cube is drawn to the total number of trials. This is also called the **experimental probability**. What would you expect the results of the experiment to be? Why? The relative frequency ratios are based on the data collected from an experiment. Compare the relative freqency ratio results with the theoretical probability ratios. Do these results differ? If so, why?

   c. Combine each group's data collected in Exercise 1b. Calculate the relative frequency ratios for the class. Compare these values with the values found in your group. Are they closer to the theoretical probabilities? Why or why not?

   d. Suppose a pink cube is added to the bag. How would this change the theoretical probabilities? How might it change the results of the experimental probabilities? Conduct the experiment. How were your results affected?

   e. Remove the pink cube, and add a second red cube to the bag. How does this change the theoretical probabilities? How could it change the results of the experiment?

   f. How does performing the experiment a large number of times affect the results?

▶ **GROUP PROJECT B**

Work in stations to investigate simple probability situations.

2. In the the following situations, you will work in stations performing experiments and collecting information. At each station, record the following information:

PORTFOLIO BUILDER

- ◆ the sample space
- ◆ the theoretical probability
- ◆ a tally of the results of the experiment
- ◆ a calculation of the relative frequency ratios
- ◆ a paragraph describing your conclusions that includes a comparison of the relative frequency and the theoretical probability

Station 1

Toss-A-Penny 1   Place a penny in a cup, shake it, and toss it onto a table or desk. Repeat these steps 10 times. Record your results on page 63 of your Portfolio Builder.

Toss-A-Penny 2   Use your results from Toss-A-Penny 1. Record the number of heads you tossed on the class chart. Keep a running total of the number of heads and the total number of tosses that everyone has made so far. Calculate the new percent of heads that have been tossed so far. Add the above information to the class chart. Continue the class graph.

Station 2   Roll-A-Number-Cube   Use a cup and one number cube. Shake the cup and roll the number cube 20 times. Record your results on page 64 of your Portfolio Builder.

Station 3   Spin-A-Spinner   Follow the directions at your station for using a paper clip spinner. The spinner at the station is like the one below.

Spin the spinner 20 times. Record your results on page 65 of your Portfolio Builder.

3. Refer to the information you gathered at the Roll-A-Number-Cube Station.

a. List the outcomes for each of the following events on page 66 of your Portfolio Builder.

PORTFOLIO BUILDER

- ◆ rolling an odd number
- ◆ rolling an even number
- ◆ rolling a number greater than 1 but less than 6
- ◆ rolling a prime number
- ◆ rolling a multiple of 3
- ◆ rolling an 8

b. Find the theoretical probability of each event listed above.

c. Find the relative frequency from the data you collected. Compare the relative frequency with the theoretical probability. How far apart are the ratios?

4. Refer to the information you gathered in the Spin-A-Spinner-Station.

a. PORTFOLIO BUILDER List the desired outcomes for each of the following events on page 66 of your Portfolio Builder.
  - ◆ spinning an even number
  - ◆ spinning an odd number

b. What is the relative frequency? What is the theoretical probability? Compare the two. How far apart are the ratios?

c. PORTFOLIO BUILDER Suppose the space labeled 5 on the spinner is divided into two equal parts, each labeled with a 5. Draw the new spinner on page 67 of your Portfolio Builder. Now what is the probability that the pointer will land on a 5? What is the probability that the pointer will land on an odd number?

d. PORTFOLIO BUILDER Now suppose that you divided the spinner as in Exercise 4c above but that one part is labeled 5 and the other is labeled 6. Draw the new spinner in your Portfolio Builder. How does $P(5)$ change? $P(\text{odd number})$? Why?

5. Today's weather forecasts may include the probability of rain or snow. Suppose these are the probabilities for rain in the next week.

PROBABILITY OF RAIN OR SNOW	
DAY	**AMOUNT**
Monday	0.1
Tuesday	80%
Wednesday	0.01
Thursday	25%
Friday	0.9

What day(s) might you plan to go to the beach or have a picnic?

▶ HOMEWORK PROJECT A

6. From the radio, television, or newspaper, find and record the chance of rain as predicted each day for a week. What is the probability of rain for each day? What is the range of values you recorded? If the prediction for one day is 30%, what ratio does this equal? What does such a prediction mean?

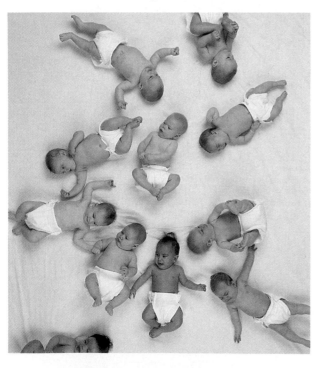

7. Suppose that a coin is used to represent the sex of an unborn child. Let heads be a girl and tails be a boy. Toss the coin 50 times and record the results. Use these results to calculate the relative frequency of boys and of girls. Is this a good model for predicting? Why or why not?

▶ GROUP PROJECT C

Analyze experiments and games of chance.

8. Norika had a cup and two chips. One chip was green and one chip was white. Norika was supposed to randomly pick a chip from the cup three times, replacing the chip after each pick. She needed to organize her work so that she could analyze her data. She started by making a list of possible outcomes.

First Pick	Second Pick	Third Pick
white	white	white
white	white	green
white	green	white
white	green	green

a. Complete Norika's list.
b. Another way Norika organized her information was to make a **tree diagram**.

First Pick	Second Pick	Third Pick

w
w ___
___
g ___
___

Complete the tree diagram.

c. What is P(2 green)? P(all green)? P(green, green, white)? P(no green)?

d. In your group, conduct Norika's experiment 20 times. Combine the data from each group. Find the relative frequency ratio for each class outcome. Compare the relative frequency ratios to the theoretical probabilities you found in Exercise 8c above.

e. Suppose one red chip is added to the cup so that Norika now has three chips. Norika randomly picks a chip three times, replacing the chip after each pick. Make a tree diagram to show the sample space for this new experiment.

f. A cup contains 12 chips. If P(white) is $\frac{2}{3}$, how many of the chips are white? What is P(not white)? If 2 more white chips are added to the cup, what is P(white) now? What is P(not white) now?

9. Suppose a family wants to have two children.

a. List all of the possible outcomes showing the gender of the children for a family that is planning to have two children.

b. What is P(both boys)? P(a boy and a girl)? P(at least one girl)?

▶ HOMEWORK PROJECT B

10. Suppose three coins are tossed.

a. How many different outcomes are possible? What is the sample space?

b. Make a tree diagram of your possible coin tosses.

c. List the probabilities for all outcomes.

11. A bag contains five cubes having the colors red, orange, yellow, and green. One cube is selected at random. Suppose P(red) is $\frac{2}{5}$ and P(orange) is $\frac{1}{5}$.

a. What is P(yellow)? P(green)? P(purple)?

b. If another red cube is added to the bag, what is the sample space? List the probabilities of picking each color.

▶ GROUP PROJECT D

Analyze experiments and games of chance.

Sometimes there are extra conditions that you need to think about before you calculate the probabilities.

12. Consider the three similar probability situations that involve tossing three coins.

a. Find P(all heads).
b. If the first toss is heads, find P(all heads).
c. If one of the coins is heads, find P(all heads).
d. What is the same or different about the three problems?
e. What conditions were added?
f. Find the probability of each situation.

13. Norika had a cup and three chips. One chip was green and two chips were white. Norika is to randomly pick a chip from the cup three times, replacing the chip after each pick.
a. Make a list of the possible outcomes.
b. Make a tree diagram of the information.
c. What is P(2 green)? P(all green)? P(white, white, green)? P(no white)?

14. Now suppose you pick a chip, *do not replace it*, and then pick a second chip.
a. How does this affect the problem? Discuss this in your group.
b. If you pick green first, what is left? If you pick white first, what is left?
c. ▪ PORTFOLIO BUILDER The tree diagram will also change. Think about the effect you discussed and complete the tree diagram like the one below on page 68 of your Portfolio Builder.

First Pick	Second Pick	Third Pick	Ordered Triples
g	w ⟶	w	(g, w, w)
	w ⟶	—	(g, w, ___)
w	— ⟶	—	(w, g, ___)
	— ⟶	—	(w, ___, ___)
w	— ⟶	—	(w, ___, ___)
	— ⟶	—	(w, ___, ___)

There are now only six possible outcomes. Notice the column on the right above. A column like this shows the order of your picks, and in this case, is called an **ordered triple** because three picks are necessary each time to empty the cup.

d. ▪ PORTFOLIO BUILDER On page 68 of your Portfolio Builder, find P(g,w,g), P(w,g,w), P(g,g,g), P(white on the first pick), and P(green on the second pick).

15. Four cards numbered 1 through 4 are placed facedown on a table to be drawn. Suppose you draw a card one at a time, without putting it back on the table.
   a. Make a tree diagram or list to show the possible orders of your draws.
   b. Next to the tree diagram, list the ordered quadruples.
   c. Find $P(1,2,3,4)$, P(last draw is a 4), P(last draw is an even number), and P(last draw is 1 or 2 or 3).

▶ **GROUP PROJECT E**

Simulate a probability experiment.

16. COMPUTER ACTIVITY The computer can be used to simulate a probability experiment using the random number function. Because the computer can complete hundreds or thousands of trials quickly, it becomes a very powerful probability tool. Refer to Exercise 13 where one green and two white chips are drawn randomly. The BASIC program below will simulate choosing a chip from the cup.

```
10 REM 1 Green, 2 White Chips
20 PRINT "Number of Trials":
   INPUT Trials
30 FOR n = 1 To Trials
40 LET x = INT(3*RND(n))
50 IF x = 0 THEN g = g + 1
60 If x = 1 OR x = 2 THEN w = w + 1
70 NEXT n
80 PRINT G/Trials, W/Trials
90 END
```

   a. Line 20 allows you to set the number of trials. In line 40, what random numbers are produced? Which color does the value of 0 represent in line 50? What meaning can you give for the ratios in line 80?
   b. Suppose you used two green chips, three white chips, and one red chip. Complete the following lines on a piece of paper to revise the program to simulate this experiment.

```
40 LET x = INT(___*RND)
50 IF x = 0 OR x = 1 THEN
   g = g + 1
55 IF x = ___ OR x = ___ OR x =
   ___ THEN w = w + 1
60 If x = 5 THEN ___ = ___ + 1
80 PRINT _____, _____, ____
```

IS IT FAIR?

Many games involve chance. Think of games you have played where a random outcome, such as rolling a number cube, is used. We could say the cube is fair if, in the long run, each number is rolled about equally often. That is, each number is rolled about $\frac{1}{6}$ of the time. Are all games of chance fair? Is a game fair if it uses random steps? Let's see how to think about fair games.

▶ GROUP PROJECT

Investigate the fairness of games.

1. In your group, play a game with two coins. Use a game board like the one shown below.

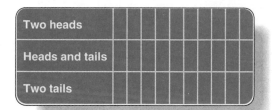

Two heads										
Heads and tails										
Two tails										

Have one group member toss the coins. The other members will each choose a row. With each toss of two coins, mark an X on the game board in the corresponding row. See who will be the first person to fill their row of 10 spaces.

   a. Play the game. Be sure to toss the two coins each time. How many groups of two tosses were made before someone filled their row completely? Find the relative frequencies for each player.

   b. Play the game three times. Record the relative frequencies for each player after each game. Explain your findings.

   c. Combine your data for the three games. Record the total number and relative frequency for each outcome. Share your results with another group. Does this game seem fair? Why or why not?

   d. What is the sample space of this game? Make a tree diagram. Use it to explain why the game turns out the way it does.

▶ PARTNER PROJECT

Investigate the fairness of games.

2. In your group, play the following game. Two people, player A and player B, will play at a time. Put one chip that is red on both sides and one chip that is red on one side and white on the other in a cup. Take turns shaking the cup and dumping out the chips.

   ◆ Player A scores a point if both chips show the red side up. Player B scores a point if one chip is red and one chip is white.

   ◆ The game is over when one player has 10 points.

   a. Do you think this game is a fair game? Why or why not? Discuss this with your group before playing.

   b. Play at least five games and record the results of each game. Calculate the relative frequency of player A scoring 10 points first, and of player B scoring 10 points first. Based upon your results, does the game appear to be fair? Why or why not?

   c. Analyze this game. List all of the possible outcomes. You may want to make a tree diagram to help you. Find the theoretical probability for player A and player B. Based on this analysis, is the game fair? Why or why not?

   d. Suppose another red-red chip is used.

   ◆ Player A scores a point if all three chips show red, otherwise player B scores the point.

   Play the game and observe the results. List all of the outcomes to find the sample space. Find the relative frequency and compare it to the theoretical probability. With your partner, discuss whether this game is fair.

   e. Think about a game with 1 red-red and 2 red-white chips.

   ◆ Player A scores a point if all three chips show red, otherwise player B scores a point.

   Write a paragraph explaining whether this version is fair.

   f. Make up your own version of a game with some combination of red-red and red-white chips. Ask another group to predict if your game is fair and then play your game. Is it a fair game? If not, how could you change the game to make it fair?

3. Play a game using 3 chips in a cup. Use a red-blue, red-white, and blue-white chip.

   ◆ Player A scores a point if all three chips are a different color.

   ◆ Player B scores a point if two chips are the same color.

   ◆ The game is over when a player scores 10 points.

   a. Predict whether this game is fair. Discuss your reasoning with your group before playing.

   b. Play the game at least five times. Record your results for each game. Calculate the relative frequency for each player. Based upon your data, does this game appear to be fair?

   c. Analyze the game by listing the sample space. How many outcomes will there be? Make a tree diagram to help determine the probability for player A and player B.

d. Is this game fair? If not, how could you change the rules to make it fair? How is this game different from the two-chip game in Exercise 2?

4. Study the grasshopper game below for two players. Make a playing board like the one shown. To begin each turn, place a chip on the home, or H, spot. Each turn consists of tossing a coin and moving the chip three times.

- If the coin lands heads, move the chip one space to the right.
- If the coin lands tails, move the chip one space to the left.
- For each turn of three tosses, player A scores a point if the chip ends up on either I spot.
- Player B scores a point if the chip ends up on the H, J, or K spots.
- A game consists of 10 turns.

a. Predict which player will win. Does this game appear to be fair? Why or why not?

b. Play the game. Record the results for each turn in a chart for each player. Based on the data, explain whether the game is fair.

c. Analyze the game by listing the sample space. Use a tree diagram to represent tossing a coin three times. How many outcomes are there? For each branch of your diagram, decide on which spot the grasshopper will end. Use your analysis to decide on the fairness of the game.

d. If the grasshopper game is not fair, how could you change the rules to make it fair?

e. Make up a different game that is based upon the grasshopper game idea. Analyze your game to decide if it is fair. Present your game as a problem for other students to play and analyze for fairness.

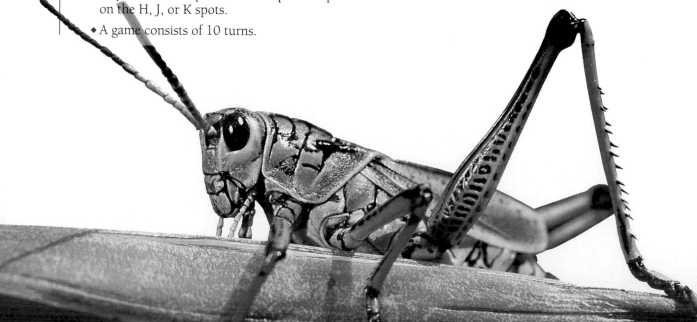

▶INDIVIDUAL PROJECT

Investigate the fairness of games.

5. Happy Hopper is a solitaire game. Imagine a Happy Hopper is stranded on a 3-by-3 square island, WXYZ. There are two bridges labeled B to safely leave the island.

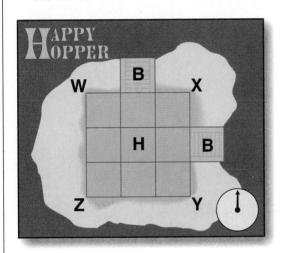

To begin the game, use a chip with an arrow marked on it to represent the Happy Hopper. Place the Happy Hopper on the home, or H, square with the arrow pointing horizontally or vertically. For each turn, toss a coin two times.

- On the first toss, if the coin lands heads, turn the arrow left, or counterclockwise, 90 degrees.
- If the coin lands tails, turn the arrow right, or clockwise, 90 degrees.
- On the second toss, if the coin lands heads, move the chip forward to the next square without turning it.
- If the coin lands tails, move the chip backward one square without turning it. Play until the Happy Hopper lands safely on a bridge or hops into the sea.

a. Does the Happy Hopper have a fair chance to escape onto a bridge?

b. Play the game at least five times. Record the sequence of squares on which the Happy Hopper lands by numbering the squares with ordered pairs. For example, the upper right corner is (1,3), and the H square is (2,2), and the upper left square in the sea is (0,1). From your data, decide if the game is fair.

c. Explain whether this game is fair.

d. Suppose the island has two more bridges attached to the squares (2,1) and (3,2). How does this change the chances of escape?

e. Investigate this situation for an island that is 5-by-5 with two bridges. How does this change the chances for the Happy Hopper?

f. For a 5-by-5 island, how many bridges would you add, and where would you place them to make a fair game?

MORE EXPERIMENTS WITH CHANCE

In this activity, you will investigate more games of chance. You will learn how to calculate the odds of winning a game.

▶ GROUP PROJECT

Explore number cube games.

1. As a class, play the following game that is similar to Bingo. Have each group member make a game board like the one below.

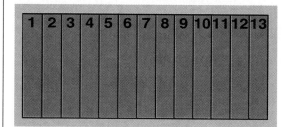

Individually, place eleven chips in any of the columns on the game board. Two number cubes will be rolled for the class and the two numbers will be called out and recorded on the chalkboard. If you have a chip on the number that is the sum of the cubes, you may remove that chip. The game ends when the first person removes all of his or her chips.

    a. Look at your results. Identify strategies for winning. Play the game again.

    b. In your group, discuss the results. Try to identify strategies for ending the game first.

    c. Play the game again but this time you cannot put more than five chips on any one number. Discuss a strategy for playing with this new

rule. Play this game at least three times. Study your results. Revise your strategy if necessary.

2. **PORTFOLIO BUILDER** Analyze how sums are found from two number cubes. How many outcomes are possible when tossing two number cubes? On page 69 of your Portfolio Builder, complete the chart like the one below showing the possible outcomes.

First Number Cube

+	1	2	3	4	5	6
1	2					
2		4				
3						
4	5					
5						
6						12

Second Number Cube (row labels 1–6)

    a. Find the probability for each sum in the chart. For example,
P(rolling a sum of 2) = ?
P(rolling a sum of 3) = ?

    b. Roll two number cubes 20 times in your group and tally the sums.

    c. Use the data to find the relative frequencies for each sum 2 through 12.

    d. Combine the data from all groups to obtain a frequency chart for the class. Compare the relative frequency with the theoretical probability of the sums of the number cubes. Then compare these with the results of the game in Exercise 1.

3. **PORTFOLIO BUILDER** In this Exercise, you will be playing a game similar to that in Exercise 1. As you go along, record your information and answer the questions on pages 70 and 71 of your Portfolio Builder.

a. Suppose the game rules are changed to use the difference of the two number cubes. Complete the chart like the one below on page 70 of your Portfolio Builder. Subtract the lesser number from the greater number.

First Number Cube

−	1	2	3	4	5	6
1	0		3			
2						
3						
4			1			
5						
6						

Second Number Cube

b. Analyze the chart to decide where you would put your chips on the game board. Record your strategies on page 70 of your Portfolio Builder.

c. If necessary, revise your game board on page 71 of your Portfolio Builder.

d. Play the game as a class. Did your strategy help you win? Why or why not? Record this information on page 71 of your Portfolio Builder.

e. Revise your strategy on page 71 of your Portfolio Builder. Place your eleven chips and play the game again.

4. **PORTFOLIO BUILDER** Refer to the game in Exercise 1. If the product of the two number cubes is used, how many columns are needed on the game board? Make a product chart on page 72 of your Portfolio Builder and revise your game board. Place your eleven chips and play the game. What strategy did you use to place your chips? Why?

5. In Unit 1, Activity 4, Exercise 13, you explored a problem by making and using two sets of colored cards numbered 1, 2, 3, 4, 5, and 6. The goal was to pair cards of each color, find the difference, and then make a product of the six differences.

a. Suppose you shuffle these cards and randomly pick one card from each set of six cards. Match the cards and find the difference. Do this six times. Form the product of the six differences. What is the probability that the product will be odd?

b. Find the probability of producing the odd number that can occur most often.

c. Reduce the two sets of cards to include only 1, 2, 3, and 4. How does this change the probability of randomly getting an odd product from the four differences?

d. Expand the two sets to include the numbers 1 through 8. What is the probability of getting an odd product from the eight differences when you randomly draw pairs?

▶ HOMEWORK PROJECT A

6. **PORTFOLIO BUILDER** Use the sum chart from page 69 of your Portfolio Builder to help you answer each of the following.

a. $P(5,5) =$

b. P(an odd sum) =

c. P(sum of 11) =

d. P(sum of either 4 or 10) =

e. P(sum is a prime number) =

f. P(rolling two different numbers) =

g. P(sum is greater than 5 but less than 10) =

h. P(sum is greater than the product of the two numbers) =

i. If sum is even, P(both numbers odd) =

7. Think of a game that you have played with a friend or relative that involves chance. What are the rules? How is chance involved? Is the game fair? Why or why not? How could you make

it more fair? Write a brief report to present to the class.

▶ PARTNER PROJECT

Explore probabilities in terms of odds.

8. The **odds** of a certain event occurring is the ratio of the likelihood it will happen to the likelihood it will not happen. When rolling a single number cube, the **odds in favor** of rolling a 3 are 1:5. Of the six possible outcomes, rolling a three is likely 1 time, while it is not likely 5 times.

a. **PORTFOLIO BUILDER** When two number cubes are rolled, what are the odds in favor of getting a sum that is less than 7? Use the sum chart on page 69 of your Portfolio Builder to help you. What are the **odds against** a sum that is greater than 5 but less than 10?

b. When two coins are tossed, what are the odds in favor of tossing two heads? Against tossing two heads?

c. Refer to the grasshopper game in Activity 3, Exercise 4. What are the odds in favor of player A? Player B?

9. A wallet contains four different bills, a $20, a $10, a $5, and a $1. You randomly pick two bills. What are the odds against your picking $15?

10. Solve these problems.

  a. Two number cubes are rolled twice. You win if both sums are 7. What are the odds that you will win? This situation can be analyzed like a coin toss, as you can win (success) or not win (failure). Make a tree diagram for the total sample space. Find the probabilities for each branch and then compute the probabilities for each outcome in the sample space.

  b. Two number cubes are rolled twice. You win if both sums are even. What are the odds that you will win?

► **HOMEWORK PROJECT B**

11. ![PORTFOLIO BUILDER] Study this difference game. Use the difference chart on page 70 of your Portfolio Builder to help you. Two number cubes will be rolled.

  • Player A scores a point if the difference is 3, 4, or 5.

  • Player B scores a point if the difference is 0, 1, or 2.

  Think about this game. Is it fair? If not, which player has the advantage? Why?

12. Study this odd and even game. Two number cubes will be rolled.

  • Player A scores a point if the sum is even.

  • Player B scores a point if the sum is odd.

  • The game is over when the first player scores 10 points.

  Without playing the game, decide whether this game is fair. If not, which player has the advantage? Why?

13. If a place kicker attempts 3 field goals in a football game, list the sample space using success and failure.

14. A batter has a batting average of .250. What is the probability that the batter will get two hits if she is at bat twice?

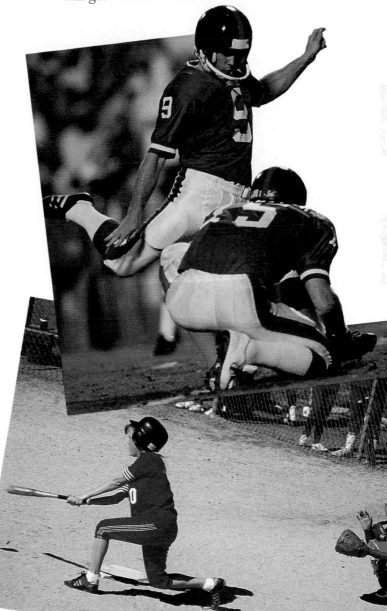

▶ INDIVIDUAL PROJECT

Simulate rolling a number cube.

15. COMPUTER ACTIVITY The computer can be used to simulate the rolling of a number cube. Study the following program.

```
10   REM Roll a Number Cube
15   PRINT "Number of Trials"
20   INPUT Trials
30   FOR n = 1 TO Trials
40   LET number cube =
     INT(6*RND(n))+1
50   LET Outcome(number cube) =
     Outcome(number cube) + 1
60   NEXT n
70   FOR row = 1 TO 6
80   PRINT Row, Outcome(row)
90   NEXT row
100  END
```

a. RUN the program. Explain how the computer is "rolling" the number cube and keeping track of the result. Based on the results, is the computer simulating a fair number cube?

b. In this activity, you have investigated rolling two number cubes. Complete these BASIC instructions to revise the program to roll two number cubes.

```
40   LET number cube 1 =
     INT(6*RND) + 1
45   LET cube 2 = ____
50   LET Outcome(___,___) =
     Outcome(___,___) + 1
60   NEXT n
70   FOR row = 1 TO 6
75   FOR col = ___ TO ___
80   PRINT Row, Col,
     Outcome(___,___)
85   NEXT col
88   PRINT
90   NEXT row
```

c. RUN the revised program. Study the output to compare the relative frequencies with the theoretical probabilities.

d. How could the program be changed to produce the sum of the two number cubes?

GIVE IT A SPIN!

You have probably seen how a circle can be used to model a fraction such as $\frac{1}{2}$ or $\frac{3}{4}$. In this activity, you will make your own spinners with specific probabilities, solve spinner puzzles, estimate and find probabilities, and find relative frequencies.

▶ GROUP PROJECT

Investigate chance with circular spinners.

1. Make a spinner where the chance of each outcome is a known probability such as $\frac{1}{3}$.

   a. Use a protractor to measure. How many degrees there are in a circle? How many degrees are in each third of the circle? Use the protractor to mark the number of degrees needed to show thirds.

   b. As you did in the Spin-A-Spinner Station, use a pencil and a paper clip to make a spinner. With your pencil tip at the center of the circle, hold the paper clip so it behaves like a pointer. You may want to open one end of the paper clip to make a pointer. Spin the paper clip to test its operation.

   c. Spin the pointer 30 times. Record each result. Compute the relative frequencies for each outcome. Compare the relative frequency with the probability.

   d. Make a circle and divide it using your protractor to show the following probabilities.

   $$P(A) = \frac{1}{2}$$
   $$P(B) = \frac{1}{6}$$
   $$P(C) = \frac{1}{3}$$

Make the circle into a spinner. Spin the spinner 42 times and record your results. Based on the probabilities, how many times would you expect the pointer to point to A? To B? To C? Compute the relative frequencies, and compare them to the probabilities.

e. Make a circle and divide it with a protractor to show the following probabilities.

$P(\text{red}) = \frac{3}{5}$

$P(\text{white}) = \frac{3}{10}$

$P(\text{blue}) = \frac{1}{10}$

Spin the spinner 30 times. Do the relative frequencies compare closely with the probabilities?

2. Solve the following spinner puzzles using the clues given. Make a sketch of each spinner.

a. **Spinner A**

◆ There are three numbers, and 4 is the greatest.

◆ One of the numbers should come up half of the time.

◆ The number 2 comes up about one fourth of the time.

◆ The most likely number to come up is 1.

◆ If you add three numbers on the spinner, you get 7.

◆ The sum from two spins might be 8, but never 7.

b. **Spinner B**

◆ The four numbers, all different, are equally likely.

◆ It is impossible to get an odd number from sums of two spins.

◆ The most likely sum of two spins is 8.

◆ The least sum from two spins is 2 and the greatest is 14.

◆ The sum from three spins is never an even number.

◆ From two spins, the sum is just as likely to be 10 or 6.

3. **PORTFOLIO BUILDER** Construct the two spinners described below on page 73 of your Portfolio Builder. Use them to investigate the sum found from spinning both.

Spinner C - divide into 5 equal parts labeled 1, 2, 3, 4, and 5.

Spinner D - divide into 2 equal parts labeled 1 and 2.

a. In your group, complete 20 trials by spinning both spinners and recording the sum on page 73 of your Portfolio Builder. Make a chart on page 74 of your Portfolio Builder to show the possible outcomes and to find the relative frequency of each outcome.

b. Combine the data from all of the groups. Compute the relative frequency of each possible sum. Make a chart of this experiment like the one you made on page 69 of your Portfolio Builder. Calculate the probability of each outcome. Compare the observed relative frequencies with the probabilities.

4. Use the two spinners, A and B, that you made for Exercise 2.

a. Spin both spinners and record the sum. What are the possible outcomes? Complete 20 trials. Find the relative frequency of each outcome.

b. Make a tree diagram or chart for this experiment. What is the probability of each sum?

5. Study the data in the frequency distribution below.

Circular Region	Frequency
A	82
B	27
C	7
D	69
E	15

Assume the data has resulted from using a spinner that has the five labeled regions. Use the data to estimate the size, in number of degrees, of each region.

6. Investigate the mystery spinner below.

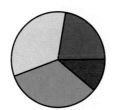

a. Make a spinner using the pattern above. Determine from experimental trials your estimate of the fractional part of the circle for each color. Spin the spinner at least 50 times. Find the relative frequency for each color. How do these frequencies permit you to estimate the fractional parts of the circle?

b. In your group, share and combine your data. Find the relative frequencies. With many more trials, should the estimates of these probabilities be improved? Why or why not?

c. Think of another way you can measure and estimate the fractional parts of the circle. Compare the results with the fractions found by the trials.

7. An unusual spinner is made using two concentric, overlapping circles. Investigate the behavior of this spinner as you use it with the Happy Hopper problem introduced in Activity 3.

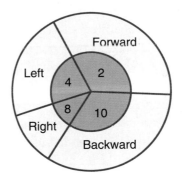

a. Use the pattern to make a spinner. Assume that the labels describe the direction the Happy Hopper moves. For the inner circle, the numbers refer to the diagonal squares at the 2 o'clock, 4 o'clock, 8 o'clock, and 10 o'clock positions.

b. Assume a 5-by-5 island with two bridges at (0,3) and (3,6). Start with Happy Hopper at (3,3) facing up, or north. For each turn, flip a coin and spin the spinner. If the coin shows heads, use the outer circle. For tails, use the inner circle. Investigate the behavior of the Happy Hopper. Record the sequence of moves, using the number pairs. Perform several trips to see how often the Hopper escapes. Look at the distribution of the Hopper ends up. How are these related to the spinner?

c. Design a similar spinner that will allow the Happy Hopper to escape about $\frac{3}{4}$ of the time. Collect data to demonstrate that your spinner works.

8. COMPUTER ACTIVITY We have seen how the circle can be used as a spinner to model probability situations. Circles can also be used as dart boards to model probabilities. Think of throwing a dart as a random action. The RND function on the computer can be used to simulate where the dart lands.

a. First, suppose a circle with a radius of 1 unit is placed in a square with sides of 4 units as shown below.

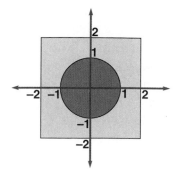

Imagine randomly throwing darts at the square. What fraction of the darts should land in the circle? Why?

b. Study this BASIC program.

```
10  REM Simulate Simple Dart Board
20  FOR Dart = 1 TO 10000
30  LET x = (5*RND(x) - 2)
40  LET y = (5*RND(y) - 2)
50  IF x^2 + y^2 ≤ 1 THEN
    Hits = Hits + 1
60  NEXT Dart
70  PRINT (Hits/10000)
80  END
```

In line 20, how many darts are thrown? In lines 30 and 40, a point (x,y) is randomly chosen to land in the square. Why do the expressions produce the desired random numbers? In line 50, if the point (x,y) is inside the circle, what happens? In line 70, what is the meaning of the ratio computed?

c. RUN the program. What do the results mean? How can you test the results by comparing to what should be expected?

d. Think about a more familiar dart board with concentric circles, such as the one pictured below.

How would you revise the computer program to simulate this dart board?

CHECKING YOUR PROGRESS

In this activity, you will demonstrate your knowledge of probability. You will apply ideas you have learned from this unit to solve situations involving chance.

This activity is a formal evaluation of your knowledge of probability. Your teacher will give you specific directions about how you should complete this activity.

PART I INDIVIDUAL ASSESSMENT

1. On your own, select and solve two of the following.
   a. A bag contains 3 white, 4 blue, and 5 red chips. One chip is drawn from the bag. Find P(blue). Find P(red). Find P(blue or red).
   b. If you randomly choose any two-digit whole number, what is the chance that the difference of the digits is less than 5?
   c. Suppose the faces of one number cube are the first six even whole numbers and the faces of a second number cube are the first six odd whole numbers. When the number cubes are rolled, what is the probability that the product will be an odd number?
   d. A deck of 52 playing cards has four suits of 13 cards each. Diamonds and hearts are red, while clubs and spades are black. If a card is drawn randomly, find P(face) and P(black). Then find P(face or black).
   e. If three coins are flipped, what is the chance of getting all heads or all tails?

2. Write a paragraph describing what you think are the most important ideas you have learned in this unit and give an example to illustrate your understanding of each.

PART II PARTNER ASSESSMENT

With your partner, solve the following problem.

3. Refer to the Marriage, Ideal Family, Rolling a Number Cube, and Baseball Batting Average problems posed in Activity 1. Select and solve one of these

problems. Write a complete report of your solution to be added to your portfolio.

PART III GROUP ASSESSMENT

Create a game.

4. As a group, invent a game that uses either cards, cubes, number cubes, or spinners to be used in a station. Set up your station that includes the necessary materials and a poster of the game rules. Turn in a complete description of your game including the rules, the mathematics used, and the fairness of the game. The groups in your class will take turns playing each others' games. You will be graded on mathematical correctness, originality, neatness, and completeness. Take a photograph of your station to be added to your portfolio.

PART IV UNIT PROBLEM

Complete your individual and group work on your unit problem. Write your report and present it to your class.

ENDING UNIT 5

Now that you have completed Unit 5, refer to page 75 of your Portfolio Builder. You will find directions for building your personal portfolio from items you made in this unit.

Size and Shape In The Plane

The guitar is a popular musical instrument, originating in Egypt over 5 000 years ago. Historians believe that the Moors brought the instrument to Spain when they invaded in 711. The guitar's design remained relatively unchanged until the mid 1800s, when a skilled carpenter, Antonio de Torres, began to make guitars in Seville, Spain.

Torres designed guitars with larger outlines, or *plantillas*. In addition, he seemed to understand basic principles of resonance. He designed his guitars to be lightweight, allowing for an improved quality in sound. It is remarkable that from the introduction of the large plantilla in the 1850s, the smaller concert guitars, prevalent in the early 1800s, became extinct.

Torres created guitars for 36 years, from 1852 until 1869 and from 1875 until 1892, the year of his death. During these years, it is estimated that he produced about 320 guitars, an average of about 10 guitars per year.

ABOUT THIS UNIT

In Unit 4, you saw how we can visualize and represent solid, or three-dimensional, shapes. Many shapes in our world are two-dimensional or flat. We say that these shapes are in a **plane**.

In this unit, you will investigate relationships involving various shapes and their parts, as well as measures of figures called perimeter, area, volume, and surface area. You will find these measures and look for patterns connecting them. In your group, you will discover patterns, formulas, and make predictions. You will make some of the most beautiful curves of mathematics and explore their properties. You will use geometric ideas to solve one of the Unit Problems posed below. You will see that geometry can be fun!

▶ UNIT PROBLEMS

The Unit Problems posed in this unit deal with housing, wealth, health, and population, all possible topics for the Course Problem. Early in this unit, you should select one of these Unit Problems.

Most of this project will be done individually, though you will use your group for ideas and reactions to your plan.

Begin work on your project as soon as possible, referring back to it as you progress through the unit. By the end of Activity 5, you should be nearly finished with your individual work. Further directions are given in Activity 7. You may wish to read those directions now.

1. HOUSING

In Unit 4, the housing Unit Problem dealt with planning a dream residence. In this Unit Problem, you can continue and extend that work. The goal is to present a detailed drawing of the interior plans.

a. Refer to the building plans you made for the exterior of your dream residence. Use these plans to develop carefully drawn-to-scale floor plans for the interior. Be sure to show where doors, windows, stairs, built-in cupboards, closets, bathroom fixtures, and kitchen counters and fixtures will be placed. You will want to determine the actual sizes of these items.

b. Assume that you are going to paint the ceilings and paint or wallpaper the walls of each room. You may need to contact a paint store to obtain color or pattern choices, prices, and rates of coverage. If you can obtain paint color strips or wallpaper samples, you might want to attach each choice to your floor plan to show how it would be painted. Compute the surface areas of walls, and use the coverage rates of the paint to estimate how many gallons of each color of paint would be needed for each room and how much the material would cost. If you are using wallpaper, estimate the number of rolls of wallpaper you would need and the cost.

c. Write and present a detailed report of your plans with cost estimates. In the summary, identify the key geometric ideas you used in your project.

2. WEALTH

For those people with sufficient wealth, travel is often a favorite leisure pastime in our modern world. Do you have a dream trip you would like to take? How about going to Australia, Hawaii, Egypt, or even Mars? Here is your chance to plan this trip. Your goal is to present a complete picture of the trip, including scale drawings showing the routes to be followed.

a. In your group, discuss your ideas for your dream trip. Help each other with ideas and reactions to clarify what is needed to document the trip. Write a description of your overall goals and itinerary for your trip.

b. Individually, develop a route and a schedule for the trip. Make rough sketches of the path for the trip.

display foods. In this project, your goal is to use geometry to understand what is done with packaging and to design an alternative package and shipping carton.

a. In your group, plan an investigation of the geometry of containers, including the packages and their shipping cartons, in your grocery store. Your overall goal is to describe the containers geometrically. In addition to rectangular boxes, include as many other shapes as you can. Write a description of the study you plan to conduct.

b. Individually, each group member will study at least 25 containers and then compile the data collected. Each member will carefully measure (in centimeters) and tabulate the dimensions of each package. Also record the count or weight of the items in the package and, where appropriate, measure the dimensions of the items inside. For these packages, ask the grocer to allow you to measure the cartons used to ship these products. Record the number of packages shipped in each carton. You may want to collect as many cartons as possible and flatten them to show the nets used to produce them.

c. Individually, analyze the data and any computed measures; for example, area, volume, surface area, or perimeter. Think about using statistics such as the mean, median,

Share these in your group to get suggestions for improvements.

c. Individually, prepare a scale drawing of the path of your trip. You may use a map to help you. Document arrangements for all transportation, including type, costs, duration, and other factors such as risks, breakdowns, cancellations, and alternatives. You may want to collect actual schedules for airlines or trains, or materials from a travel agent including photos to portray the sites you want to visit.

d. Write a complete report of your dream trip, featuring geometric drawings and graphs with data charts. In the summary, identify the key geometric ideas you used in your project.

3. HEALTH

Our grocery stores are filled with hundreds of different containers used to package and

mode, or range, and graphs to present the results of your study. Make scale drawings of the packages in your sample. Select one of the products and develop an alternative package and rationale for marketing it.

d. Write a report and present your results. In the summary, identify the key geometric ideas you used in your project, including area, perimeter, surface area, maximum or minimum, and properties of polygons and curves.

4. POPULATION

In many heavily-populated areas such as large cities, geometry has been used to plan streets and roads, subway and train systems, freeways and highways, and bus routes.

a. Investigate the design of these various systems for transporting people in your area. What is the geometric pattern for the streets and roads? Why? What geometry can be found in the layout of freeways and other highways? What are some possible factors in determining the paths used? How much land area is used for streets and roads? For freeways and highways? Find the rate of usage (for example, number of people per day) for each type of transportation system in your area. What can you conclude? How might it be improved? Explain.

b. Study a map of a city such as London, New York, or Washington D.C. Locate the path of the subway system. What geometric patterns can you find? How much area in the city is used by the subway system? What is the average daily rate of usage? What percentage of the population uses the subway? How could the system be improved? Explain.

c. Study the freeway and highway system in a metropolitan area such as Los Angeles, New York, or Chicago. What geometric patterns can you find? How much land area is used? What is the average daily rate of usage? What percentage of the population uses the system? How might it be improved? Explain.

THE SHAPE OF IT

In this activity, you will explore lines, angles, and other common geometric shapes. You will draw some polygons and investigate some of their properties. If you come across any terms that you are not familiar with, look them up in the glossary.

▶ GROUP PROJECT A

Use paper folding to make basic geometric shapes and investigate their properties.

An interesting and easy way to make geometric shapes is by folding paper. Use sheets of $8\frac{1}{2}$" x 11" paper. To conserve paper, make as many shapes on each sheet of paper as possible.

1. Use paper folding to make an example of each of the shapes described. As each shape is made, write the word or words on the shape to name it. For example, notice that when you fold and crease your paper, you model a line segment.

line segment

In your group, compare what you make with what others make. How do the shapes vary? Be sure that you can explain the steps needed to make the shape. For each, look around the room to find an example of the shape.

a. **Two intersecting lines**

   How many angles are formed? What seems to be true about the size of the opposite pairs of angles? Use your protractor to measure them. Compare your measures with the measures found by the other members of your group.

b. **Angle bisector**

   An **angle bisector** is a ray or a line that divides an angle in half. How do you know you have folded correctly to find the bisector of an angle? Mark three points along the bisector. Measure the distance from each point to the sides of the angle. What appears to be true?

c. **Two perpendicular segments**

   How do you know they are perpendicular? Use your protractor to measure them. What is true about the size of the angles?

d. **Segment bisector (perpendicular)**

   How do you know you have found the midpoint of the segment? Think about your folding actions. How are these similar or different from those you used to find the angle bisector? Mark several points along the perpendicular bisector. For each point, measure how far it is to the

endpoints of the segment. What can you conclude?

e. **Two parallel line segments**

   Discuss how you know they are parallel.

f. **A square**

   There is more than one way to fold a square. Compare your squares with your group members. Discuss the different ways you folded your squares.

g. **A parallelogram**

   In your group, check to see what variety of parallelograms have been made. Is a rectangle a parallelogram? Did anyone make a rhombus? What makes a rhombus special?

h. **A trapezoid**

   Compare your trapezoid with those in your group to see the variety. Now make an isosceles trapezoid.

i. **A triangle**

   After everyone has made a triangle, compare them with others in your group to see the variety of triangles made. Find examples of an acute triangle, a right triangle, and an obtuse triangle. If you do not find these in your group, make one. Now fold your paper into an isosceles triangle, a scalene triangle, and an equilateral triangle. With a protractor, measure each angle. What is the sum of the measures of the angles in each of your triangles? Compare your answers with the class. What did you find?

▶ GROUP PROJECT B

Use a geoboard to represent geometric shapes.

2. Another interesting way to represent geometric shapes involves using a geoboard and geobands. As with paper folding, the main advantage of using a geoboard is that it allows you to visualize a wide variety of polygons and makes measuring their areas easy. If a geoboard is not available, you may use geoboard dot paper. In your group, model each of the following shapes on your geoboard. As you make each polygon, compare it with the polygons made by your group members. How are the polygons alike? How are they different?

   a. each of the shapes from Exercise 1
   b. a four-sided polygon with no parallel sides
   c. a four-sided polygon with opposite sides parallel
   d. a four-sided polygon with only two sides parallel
   e. a quadrilateral with exactly one pair of perpendicular sides
   f. an octagon
   g. a right triangle
   h. an isosceles triangle
   i. an obtuse triangle
   j. an equilateral triangle
   k. a shape that touches five pegs
   l. a shape that has three pegs inside
   m. a shape that touches five pegs with three pegs inside

3. Study the following puzzle. Suppose there are three towns, A, B, and C, on a straight road. Town B is 6 km from town A and 3 km from town C. There is another town, X, on this road that is also twice as far from A as it is from C. How many kilometers is town X from town C?

   a. Model this situation on geoboard dot paper. Solve the puzzle. Is there only one solution? Explain.

b. Make up another puzzle that is similar to this one. Ask another student to solve it.

4. 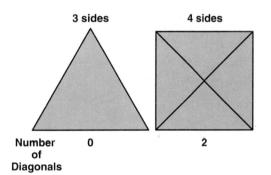 As a group, organize and conduct an investigation to answer the following question. If a polygon has n sides, how many diagonals does it have? Systematically look at each type of polygon starting with a polygon of 3 sides.

3 sides   **4 sides**

Number of Diagonals   0   2

a. For each polygon, record your findings in the chart on page 76 of your Portfolio Builder.

b. Look for and record any patterns you find in the data on page 77 of your Portfolio Builder. Then plot the number pairs to make a graph of the data.

c. Find the number of diagonals for a 100-sided polygon. Explain your work.

d. Continue the chart on page 76 of your Portfolio Builder to include an **n-gon**. An n-gon is a polygon with n sides.

5. Think about the Venn diagrams you made in Unit 1. You can use Venn diagrams to classify different kinds of polygons. Think about all quadrilaterals.

a. On a piece of paper, make a list of all of the kinds of quadrilaterals we have discussed so far. Cut apart your list so that each kind of quadrilateral is on a different slip of paper.

b. Make a Venn diagram. Sort and classify your list of all of the types of quadrilaterals to show the group to which they belong. For example, is every square a rectangle? A parallelogram? A rhombus? A quadrilateral? A trapezoid? Record your final Venn diagram on page 78 of your Portfolio Builder.

▶ HOMEWORK PROJECT

6. At one of the latest peace summits, 30 world leaders met. They sat at a very large, round table. They entered the room, took their appointed seats and shook hands with the person to their immediate right and their immediate left. At the end of the meeting, each person shook hands with all of the people they did not shake hands with at the beginning of the meeting. How many handshakes were there in all? Explain how this problem relates to something learned in Activity 1.

AREA OF SHAPES

We often need to describe the size of a geometric shape. To do so, we need to measure. In this lesson, you will use geoboard dot paper to explore how to measure the perimeter and area of various shapes. You will investigate how these two measures might be related.

▶ GROUP PROJECT A

Find the perimeter and area of rectangles.

You may recall from Unit 4 that the perimeter of a polygon is the distance around it. The area is the number of square units contained in the interior of the shape.

1. In your group, use geoboard dot paper to review the area and perimeter of rectangles. Assume that a unit of length is the distance between adjacent dots along a row or column. A square unit of area is a square made with four adjacent dots.

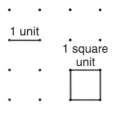

For example, in the rectangle below, the rectangle is divided into 12 square units. The perimeter of this rectangle is 14 units.

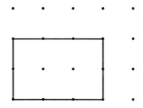

a. Find the perimeter and area of the rectangle below.

b. Use geoboard dot paper to draw rectangles with areas of 3 units and areas of 8 units. Find the perimeter of each.

c. **PORTFOLIO BUILDER** Cut out one of the rectangles that you drew in Exercise 1b. Glue or tape your rectangle on page 79 of your Portfolio Builder. Then, in your own words, explain how to find the perimeter and area of a rectangle. Write **formulas** for finding the perimeter and area of a rectangle.

▶ GROUP PROJECT B

Find the area of triangles and parallelograms.

2. Look at the right triangle below. A rectangle is drawn that includes the triangle by placing the vertices of the triangle along the sides of the rectangle as shown.

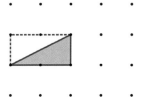

a. How would you find the area of the triangle? How does its area compare to the rectangle? Find the area of this rectangle. Then find the area of the right triangle.

b. Find the area of the triangle below.

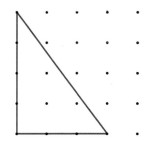

c. Find the area of the triangle below.

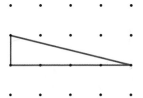

d. Draw right triangles on your geoboard dot paper that have areas of 2, 5, and 8 units.

e. 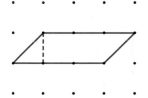 Cut out one of the triangles that you drew in **PORTFOLIO BUILDER** Exercise 2d. Glue or tape your triangle on page 80 of your Portfolio Builder. Then, in your own words, explain how to find the area of a right triangle.

3. On your geoboard dot paper, draw a parallelogram like the one below.

a. Cut out your parallelogram. What is the area of this parallelogram? Use what you learned about finding the area of a right triangle to help you. Describe how you found this answer.

b. The dotted line is called the height of the parallelogram. How does the length of this dotted line compare to the width of the parallelogram?

c. With a pair of scissors, neatly cut along the dotted line and cut out the right triangle on the left. Place this triangle on the right side of the polygon. What common shape do you have now? What is the area of this common shape?

d. On your geoboard dot paper, draw a parallelogram like the one below.

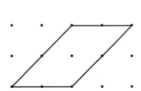

Cut out your parallelogram and find its area. Where would you cut the parallelogram so you could piece it together to form a rectangle? How does the area of the parallelogram compare to the area of the rectangle formed?

e. What is the formula for finding the area of a rectangle?

f. How can you use this formula to help you write a new formula for finding the area of any parallelogram?

g. **PORTFOLIO BUILDER** Glue or tape your parallelogram on page 81 of your Portfolio Builder. Then, in your own words, explain how to find the area of a parallelogram. Write a formula for finding the area of a parallelogram.

4. Study the examples showing families of parallelograms.

A family of parallelograms is a group of parallelograms that have some characteristics in common.

a. In the family of parallelograms below, list the characteristics they have in common.

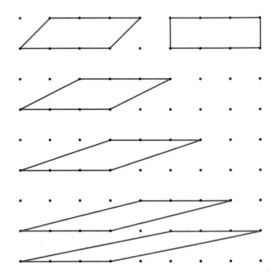

b. Copy and find the areas of each of the parallelograms above. Compare the areas. Write a conclusion to compare this family of parallelograms.

c. Make another family of parallelograms. Find the areas. Make your conclusions from Exercise 4b with this new family. What can you say about families of parallelograms?

5. Look at the triangles below.

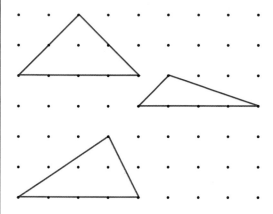

a. What kind of triangle is each? Like parallelograms, some triangles can be cut and pieced together to make a rectangle. Other triangles can be cut so that each could be half of a rectangle, or a right triangle. Find the area of each triangle.

b. 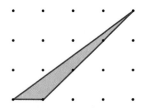 Copy the acute triangle above onto another piece of geoboard dot paper. Cut out the triangle and glue or tape it onto page 80 of your Portfolio Builder. Then, in your own words, explain how to find the area of this triangle.

6. The triangle below is an example of an obtuse triangle.

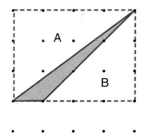

a. Explain whether or not obtuse triangles can have more than one obtuse angle. Why or why not?

b. To find the area of obtuse triangles, we can first form a rectangle that includes the triangle. Place the vertices of the triangle on the sides of the rectangle.

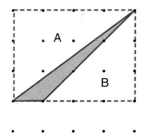

c. Find the area of the rectangle above. Subtract the area of triangles A and B. What is left?

d. Find the shaded area of the triangle below.

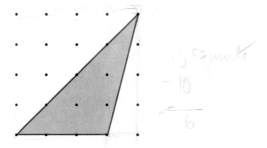

Explain the method you used. Will the same method work for other triangles? Go back to Exercises 2 and 5. Check your answers using this method.

7. Copy the triangle in Exercise 6d onto geoboard dot paper. **PORTFOLIO BUILDER** Cut and glue or tape this triangle onto page 80 of your Portfolio Builder. In your own words, explain how to find the area of this triangle.

8. Look at the family of triangles below.

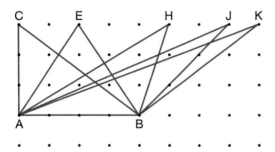

a. What characteristics do they have in common?

b. 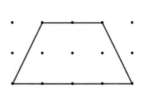 Draw each triangle by itself on page 82 of your **PORTFOLIO BUILDER** Portfolio Builder.

c. Find the area of each triangle. What conclusions can you make about this family? Make another family of triangles. What can you say about families of triangles?

d. On page 80 of your Portfolio Builder, write a **PORTFOLIO BUILDER** formula or rule for finding the area of any triangle.

▶ **GROUP PROJECT C**

Find the area of trapezoids.

9. Look at the shape below.

a. What shape is this? Use geoboard dot paper to **PORTFOLIO BUILDER** draw and cut out two of these. Put the two trapezoids together to form a parallelogram. Look back at page 81 of your Portfolio Builder. What was the formula you developed for finding the area of the parallelogram? How does the area of this trapezoid compare to the area of this parallelogram?

b. Find the area of the trapezoid below.

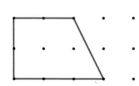

c. Find the area of the trapezoid below.

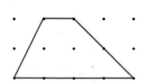

d. Glue or tape one of your trapezoids onto page 83 of your Portfolio Builder. Use the areas of the three trapezoids above and the area formula for a parallelogram to discover a formula for the area of a trapezoid. In your own words, explain how you found the area of your trapezoid. Then write a formula for finding the area of a trapezoid.

▶ HOMEWORK PROJECT

10. Look at the family of trapezoids below. What characteristics do they have in common? Find the area of each trapezoid. What conclusions can you make about the family of trapezoids?

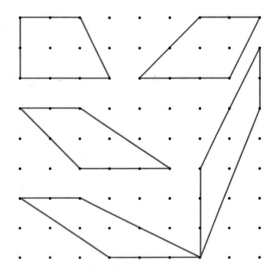

▶ GROUP PROJECT D

Find the area of any polygon.

11. Look at the polygons below.

 Turn to page 84 of your Portfolio Builder. Determine a strategy for finding the area of these shapes. Then, find the area of each polygon. Show your work.

12. COMPUTER ACTIVITY In Unit 5, Activity 5, Exercise 8, a BASIC program was given for simulating a dart board. This idea can be used to estimate the area of any region in the plane.
  a. Take a simple case: find the area of a 12-by-12 square target that is

centered in a 20-by-20 square dart board. Study the graph and program below.

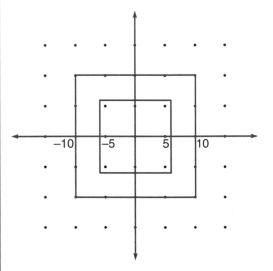

10 REM Simulate Dart Board To Estimate Area

20 FOR Dart = 1 TO 10000

30 LET x = (21*RND(x)-10)

40 LET y = (21*RND(y)-10)

50 IF x 6 AND y 6 AND x -6 AND y -6 THEN Hits = Hits + 1

60 NEXT Dart

70 PRINT "Estimated Area Is"; (Hits/10000)*400

80 END

In line 20, how many darts are thrown? In lines 30 and 40, these expressions produce the random position for the dart. In line 50, if the point is inside the inner square, what happens? In line 70, what is the meaning of the ratio? Why is the ratio multiplied by 400?

b. RUN the program. What do the results mean? What would be expected? How could you improve the estimate?

c. Change line 50 to the following:
50 IF y x + 6 AND y - x + 6 AND y - x - 6 AND y x - 6 THEN Hits = Hits + 1
Use the inequalities to make a graph of the new target. Find the expected area of this shape. RUN the program five times and record each estimated area. Find the average and compare it to the expected value.

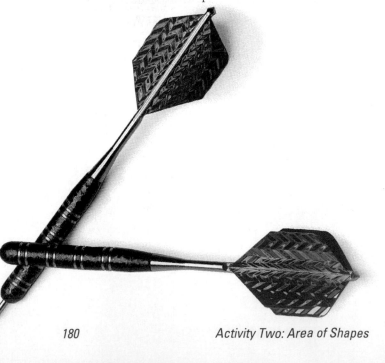

GETTING IT FENCED IN

In this activity, you will explore the connection between perimeter and area. Given a perimeter, you will find possible areas. You will see how the perimeter changes as the area is increased. You will also find the maximum or minimum area and perimeter to solve problems.

▶ GROUP PROJECT A

Work in stations to find perimeter and area.

1. In the following stations, you will be exploring perimeter and area. Record all of your information from each station neatly as it will be added to your portfolio. Be sure to label and record each station's information on a separate piece of paper.

Station 1   Building Larger Squares

In this station, you will find the area and perimeter of successively larger squares.

♦ Start with one square tile. The length of the tile is one unit by one unit.

♦ Make a table like the one shown below. Complete the table for each successively larger square.

Number of Tiles Used	Length of One Side	Area	Perimeter
1			

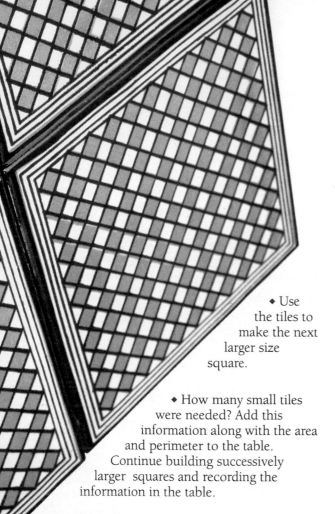

◆ Cut one of the squares along the lines as shown below to make two different-sized pieces.

◆ Tape the two new pieces together to make an irregular shape. Tape this new shape onto another piece of paper.

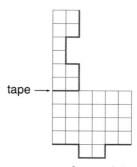

tape →

◆ Find the perimeter of your shape. Write the perimeter inside of the shape.

◆ Repeat these steps with the four other squares. Be sure to make each into a different irregular shape.

◆ How do the five areas compare now? How do the five perimeters compare? Write a paragraph to explain what you have discovered.

◆ Remove the shape that has the greatest perimeter and tape it to the class chart.

◆ Use the tiles to make the next larger size square.

◆ How many small tiles were needed? Add this information along with the area and perimeter to the table. Continue building successively larger squares and recording the information in the table.

◆ Describe any patterns you see. Compare the number of tiles used, the area, and the perimeter.

◆ What would be the area of a large square made up of 100 small tiles? Explain. What would be the perimeter?

◆ Complete the table if x was the length of one side.

Station 2   Looking At The Same Area

◆ Cut grid paper into five 6 x 6 squares. How do the areas of these five squares compare?

Station 3   Looking At The Same Perimeter

◆ Draw five different polygons on grid paper so that they all have a perimeter of 30 units.

◆ Find the area of each polygon and write this number inside the shape. Circle the polygons that have the least and greatest areas.

◆ Why do you suppose the area changes even though the perimeter remains the same? Write a paragraph to explain your conclusions.

Station 4 Shoe Area And Perimeter

◆ On grid paper, trace around your shoe. Using the grid paper, find the approximate area of the sole of your shoe.

◆ Use string to help you find the perimeter of the sole of your shoe. Cut two pieces of string that measure this perimeter.

◆ Record your name, the area, and the perimeter of your shoe on a class chart.

◆ Take one of your pieces of string and form a square. Tape this string square to a piece of grid paper. Find the area of the square.

◆ How does the area of the string square compare to the area of the sole of your shoe? Explain.

◆ Take your second piece of string and cut it in half. Make a new string square with this half and tape it to your grid paper. Find the area of this square.

◆ Is the new area one-half of the larger string area? Why or why not?

◆ Look back at the class chart. Does anyone have your same shoe area or perimeter? If the areas are the same, are the perimeters the same? If the perimeters are the same, are the areas the same? Explain.

Station 5 Hand Area And Perimeter

◆ On grid paper, carefully trace around your hand. Place your hand palm down with your fingers spread apart.

◆ Trace your hand on graph paper again with your hand palm down and your fingers together.

◆ In your group, discuss how you might measure the perimeter and area of each tracing. Think about using a piece of string to help measure the perimeter. You can use the grid on the graph paper to help you measure the area.

◆ With a partner, find the perimeter and area of each tracing of your hand. Record this data for all group members on the class chart. What can you conclude from the data?

▶ GROUP PROJECT B

Investigate possible perimeters for a given area.

2. In your group, use square tiles, geoboards, or dot paper to explore the following situation.

   Leticia plans to make a pen for her pet rabbit. She has decided that her rabbit needs a pen with an area of 48 square feet. Since she has a limited budget, she wants the length of fence to be the minimum amount possible. What shape would be best for the rabbit pen?

   a. Try different arrangements of the 48 tiles to find a minimum perimeter. Record your solutions. Be ready to explain why you know you have the minimum length of fence. What is the minimum length?

   b. Suppose that Leticia changed her mind and decided to make the perimeter a maximum for the area of 48 square feet. What shape gives the maximum perimeter?

3. Organize an investigation to find the shapes and their minimum perimeters for various areas from 1 to 10 square units.

   a. On page 85 of your Portfolio Builder, fill in the chart where you systematically vary the area and record the shape and minimum perimeter. Study the data. Look for patterns. Graph the data on page 86 of your Portfolio Builder. What conclusions can you make? If the

shape is a rectangle, explain how to find the perimeter without counting each side of the tile.

   b. Investigate the relationship of the area and the maximum perimeter. On page 85 of your Portfolio Builder, continue the chart, record the shapes and maximum perimeters. On page 86 of your Portfolio Builder, graph this data using a different color of pen. State any conclusions.

4. Your club is planning to have a dinner party. There are 12 small square tables that can be put together to form a large dining table. Assume that on one side of the small square table there will be space for one person and that the complete sides of the tables must be touching.

   a. Draw all possible combinations of tables. Find the number of people that can sit at each of these tables.

   b. What is the minimum number of people who can be seated?

   c. What is the maximum number of people who can be seated?

▶ HOMEWORK PROJECT A

5. Discuss with an adult the advantages and disadvantages of arranging the tables in Exercise 4. Make a list of these advantages and disadvantages. Which table arrangement do you like best? Why?

6. If there were 100 small square tables, what would be the minimum number of people that could be seated? What would be the maximum number? Explain.

▶ GROUP PROJECT C

Investigate how the perimeter changes when the area is increased by one unit.

7. In your group, discuss what you think happens to the area of a shape when you add another tile (one square unit).

   a. If you add a tile to a shape, does the perimeter always increase? Can it stay the same? Can the perimeter decrease when the area increases by one? Make some conjectures.

   b. Study the examples below that are made using 8 tiles. Be alert to shapes with holes and alleys.

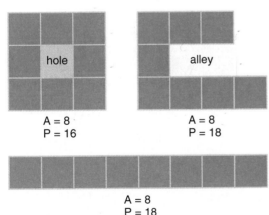

A = 8
P = 16

A = 8
P = 18

A = 8
P = 18

A **hole** in a shape is where some tiles are left out of the middle. The perimeter of the shape shown at the top left is 16 units because the inside

of the perimeter is counted as well as the outside perimeter. An **alley** in a shape is where the tiles form a kind of alleyway. The perimeter of the alley shape above is 18 units.

   c. On grid paper, draw the other possible shapes you can make with 8 tiles. Below each shape, record its perimeter and area.

   d. Now, try adding one tile to each shape in various ways. In each case, see what effect this has on the perimeter. Try to find at least one shape for each of these possible effects on the perimeter: stays the same, increases 1, 2, 3, or 4 units, decreases 1, 2, 3, or 4 units. Record the new shape on your grid paper or write "not possible".

   e. Write an explanation of the possible effect on the perimeter of increasing the area by one square unit. Can you increase the perimeter by an even number? Why or why not? Can you increase the perimeter by an odd number? Why or why not?

▶ HOMEWORK PROJECT B

8. You have probably seen or played a game that uses dominoes. Each playing piece consists of two squares joined together to share a side.

   a. A **triomino** is a pattern made by joining exactly three squares together so they share sides. You can model these three-square shapes on geoboard dot paper or grid paper.

Make all of the possible arrangements of triominoes. Check to be sure that no two triominoes are the same. How many are there?

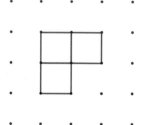

b. Repeat Exercise 6a using four squares that are joined together to share sides called **tetrominoes**.

c. Now, repeat Exercise 6a using five squares that are joined together to share sides called **pentominoes**.

d. Repeat Exercise 6a using six squares that are joined together to share sides called **hexominoes**.

e. Make a chart to organize all of your information.

▶ GROUP PROJECT D

Investigate the possible areas for a given perimeter.

9. In your group, explore the following situation.

> Luis has shopped for fencing to build a rabbit pen. He finds that he can afford a roll of fencing that is 48 feet long. So, he decides to make a rectangle that has the greatest area. Find the dimensions for the maximum area with a perimeter of 48 feet. What will the area be?

a. In your group, organize an investigation to solve this problem. Use tiles, a geoboard, or dot paper to find possible rectangles with a perimeter of 48 units. Make a chart to record your findings. What rectangle gives the minimum area? The maximum area? What are its dimensions?

b. Make a graph of the data for the perimeter of 48 feet. Select one of the sides. Be sure to record the same side each time. Plot the ordered pairs (side length, area). Connect the points. What is the shape of the graph?

c. On the graph, find the two places having an area of 120 square feet. How are these side lengths different from those you plotted before? What are the dimensions of this rectangle that has a perimeter of 48 feet? How could you use the graph to find the maximum area? Then, where on the graph would you find the dimensions of the rectangle?

d. Think about Luis' situation if other lengths of fencing could be used. For example, what would give you the maximum area for 64 feet of fencing? 100 feet of fence? What seems to be the size and shape of the solution?

e. Make graphs for these perimeters: 12 feet, 20 feet, 32 feet, 64 feet, 100 feet 30 feet, and 58 feet. In each case, find the maximum area and its dimensions. What pattern occurs?

10. COMPUTER ACTIVITY With a simple BASIC program, we can get the computer to help us find the maximum area of a rectangle for a given perimeter.

a. Complete the following program.

10 REM Rectangle-Maximum Area, Fixed Perimeter

20 REM l = Length, w = width

30 PRINT "What is the perimeter?": INPUT p

40 FOR l = 1 TO_____

50 LET w = _____ - length

60 PRINT length, width, _____

70 NEXT length

80 END

b. RUN the program. Try each of the suggested perimeters in Exercise 9e.

Does the program produce correct results? Can you find the maximum area in the output?

c. RUN the program using a perimeter of 7. Can you find the maximum area in the output? Since it appears to occur between two values, how can you change the program to get more refined results? Hint: Try using a smaller step in line 40.

► HOMEWORK PROJECT C

11. Suppose there is 200 feet of fencing to make two identical rectangular pens along already existing fencing.

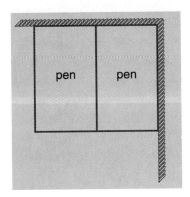

a. Investigate the possible dimensions of each pen. Then find the area of each of these pens.

b. Which pen has the greatest area?

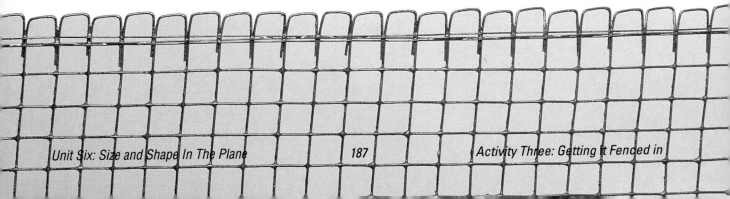

ROUND AND ROUND

What's the path of a person riding on a merry-go-round or a Ferris wheel? Up until now, you have investigated polygons and polyhedra that have straight sides and edges. Many interesting geometric shapes involve sets of points that do not lie in a straight line. In this activity, you will explore the properties of circles and learn to find the area and circumference of circles.

▶ **PARTNER PROJECT A**

Find the circumference of a circle.

In Activity 2, you found the perimeter of a rectangle. The perimeter of a circle is called the **circumference**. The **diameter** of a circle is a line segment that can be drawn from a point on the circle through the center of the circle to another point on the circle. The **radius** of a circle is the line segment that can be drawn from the center of the circle to a point on the circle. How

does the length of the radius compare to the length of the diameter of a circle?

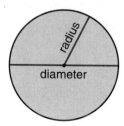

1. How many diameters can you draw in a circle?
   a. Make a three-inch square. Fold the square to find its center. Make a hole in the center of the square. Label two opposite vertices of the square A and C.

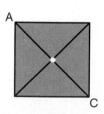

b. Place this square onto another piece of paper. Hold the square with a pencil or compass point in the center. Mark where the vertices A and C are located on the paper.

c. Rotate the square slightly. Mark where the vertices A and C are now located.

d. Repeat this process 10 times, or until the marks appear to make a circle. Connect the corresponding points with a line segment that goes through the center. What seems to be true about the number of diameters in a circle?

2. Look at the circle below. The measure of the circumference is shown by placing a tape measure around the outside edge of the circle.

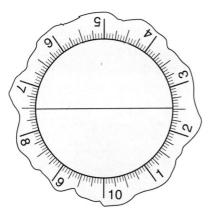

a. How could you use a piece of string to find this measure?

b. Use a piece of string and a ruler to measure the circumference of the circle above. Did you get the same measure?

3. Trace around and measure the diameter and circumference of circular objects of different sizes. You may want to take three measures of the circumference and average them. Record your findings in a table like the one below.

Diameter (d) (centimeters)	Average Circumference (C)	Ratio $\frac{C}{d}$

4. Find the ratio of the circumference to the diameter for each of the circles in Exercise 3 above. Add this to your table.

a. What do you notice about these ratios?

b. Find the symbol π on your calculator and enter it. How does this number compare to the ratios you found in your table?

5. Graph your findings from your table. Put the data for the diameter on the horizontal axis and the corresponding data for the circumference on the vertical axis. Connect the points. What shape is your graph? Explain.

6. In Exercise 4, you found that circumference divided by diameter approximates pi ($\frac{C}{d} = \pi$).

a. Explain how knowing the number π and the diameter of a circle can help you find the circumference of the circle.

b.  On page 87 of your Portfolio Builder, explain in your own words how to find the circumference of a circle. Then write a formula for finding the circumference of any circle.

c. If the circumference of a circle equals 100 centimeters, find the diameter and the radius of this circle.

d. If the diameter of a circle is 42 feet, find the radius and the circumference of this circle.

7. The wheel on Jorge's bicycle has a radius of 12 inches. How many revolutions will the wheel make while Jorge travels 500 feet?

8. A Ferris wheel is 45 feet tall. How far would you travel on this ride if you went around 10 times?

9. In Unit 3, you found that the distance around Earth at the equator is about 40 500 kilometers. Find the approximate diameter of Earth.

▶ GROUP PROJECT A

Use circumference to solve problems.

10. Suppose you had a small plate and a bicycle tire. The diameter of the plate was 5 inches, and the diameter of the tire was 30 inches.

   a. Compute the circumference of both the plate and the bicycle tire.

   b. Pretend that you had strings of these two lengths. Now suppose you added 9.42 inches to each string. If you put these new strings around the plate and the tire, how far outside each would the string be? What would be the new diameter of these string circles?

▶ PARTNER PROJECT B

Find the area of a circle.

11. Look at the circle below that has been traced on graph paper. Decide how to count squares to estimate the number of square units there are inside the circle. Share your findings with the class.

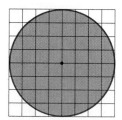

12. With your partner, use three different-sized round cans. One can should be small, another medium, and the third large. Trace around the base of each can

onto graph paper. As a class, decide on a method to count squares. Use this method to count the squares and record your estimates. Compare your findings with your group.

13. Use the same three cans as in Exercise 12.
    a. Trace the bases of the cans onto stiff blank paper such as construction paper.
    b. Measure and record the radius of each circle. Then measure and record the circumference of each circle.
    c. Divide and cut each circle into sixteen equal parts as shown below.

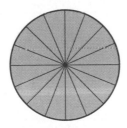

    Rearrange the pieces of each to form three separate parallelograms. Let the longest side of the parallelogram be the base. Glue or tape these parallelograms onto another piece of paper.
    d. Measure and record the height of each parallelogram. How does the height compare to the radius of the circle?
    e. Measure and record the base of each parallelogram. How does this measurement compare to the circumference of the original circle?

f. Compare the estimate of the area you made from counting the squares to the one you made by cutting and gluing the parallelograms. Which one do you think is the most accurate? Explain.

14. **PORTFOLIO BUILDER** Cut out one of your taped parallelograms from Exercise 13. Glue or tape it to page 87 of your Portfolio Builder. In your own words, explain how to find the area of a circle. Then develop a formula or rule for finding the area of any circle. Think of the rule or formula you developed for finding the circumference of any circle.

▶ HOMEWORK PROJECT

15. Suppose Luis decided that he should not limit the possible shape to a rectangle. Consider any geometric shape. With 48 feet of fencing, what shape would have a maximum area greater than what you found in Activity 3, Exercise 9? Does this seem to hold for other perimeters also? Explain your answer.

16. The radius of circle A below is 1 inch and the radius of circle B is 3 inches.

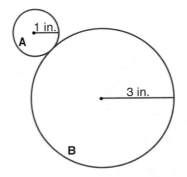

Suppose circle A starts in the position shown and rolls around circle B. How far must the center of circle A travel before it will first reach its starting position? At the end of how many revolutions of circle A will the center of circle A first reach its starting point?

17. Suppose there was a steel band fitting tightly around the equator of Earth. Suppose that you remove the band and cut it at one place, then add in an additional piece 10 feet long so that the new band is 10 feet longer than the original band. If you put the band back on the equator, how far above Earth's surface would it be? The circumference of Earth is about 24 900 miles. First make an estimate, then solve.

▶ GROUP PROJECT B

Use the area of circles to develop a strategy.

18. Look at the figures below. Suppose each of these figures represent targets.

You are going to play a game called "place the dot on the target." Each person in the group will take turns being blindfolded. The blindfolded person will be given a felt-tipped marker and be told to mark a dot onto a target of their choice. The group members will help direct the blindfolded person toward the target. Each group member will take a turn using a different color of marker. Points are earned as follows:

◆ You score 1 point if your mark is inside a circle.

◆ You score 0 points if your mark is inside a square but outside a circle.

a. Decide which target you will choose to mark.

b. Play the game. Compare the results with the results of the other groups.

c. After playing the game and comparing the results, would you choose a different target to mark? Why or why not?

d. Explain how you can use mathematics to help you make a good choice of target.

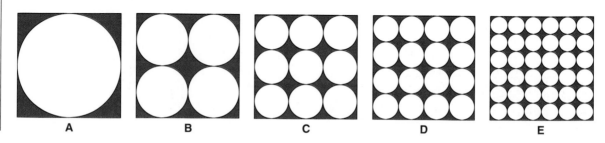

A   B   C   D   E

EXTENDING YOUR THINKING ABOUT CIRCLES TO THREE DIMENSIONS

In Activity 4, you looked at circles. In this activity, you will look at circles that are part of three-dimensional objects. You will also find the volume and surface area of cylinders.

▶ GROUP PROJECT A

Relate the volumes of right prisms to cylinders.

1. In Unit 4, you explored the volume of rectangular prisms. Draw a 3-by-4-by-2 rectangular prism.
   a. Find the volume of this prism.
   b. What is the area of the base?
   c. How can knowing the height and the area of the base help you calculate the volume of the prism?

2. Ashley has a cereal box with a base area of 24 square inches and a height of 14 inches. What is the volume of the cereal box?

3. Drew has an octagonal-shaped fish tank with a base as shown below and a height of 20 centimeters.

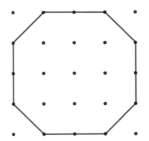

The volume of any right prism is found by multiplying the area of the base and the height. How many cubic centimeters of water will fill Drew's tank?

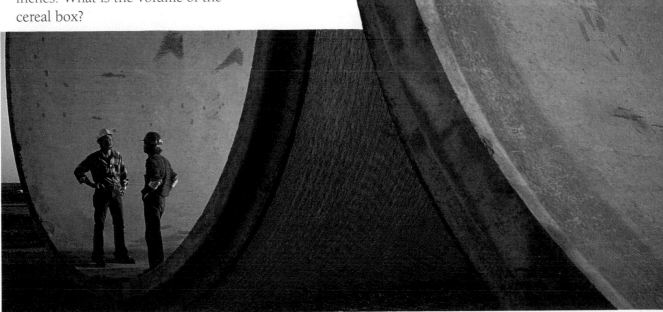

4. The volume of right prisms that equals the area of the base times the height ($V = Bh$) can be extended to right cylinders. Dane fills an empty coffee can with sand from the beach. If the base of the coffee can has a radius of $2\frac{1}{2}$ inches and the height of the coffee can is $6\frac{1}{2}$ inches, find the volume of the coffee can.

5. Dane wants to pour his sand out of the coffee can into a right rectangular prism. What are the possible dimensions of the minimum-sized box that would hold this sand?

▶ GROUP PROJECT B

Compare the volumes of cylinders.

6. Use two pieces of $8\frac{1}{2}$" x 11" paper. Roll one piece of paper into the shape of a cylinder and tape it so that the height is the $8\frac{1}{2}$" width of the paper.
   a. Take the second piece of paper and roll it into the shape of a cylinder and tape it the other way so that the height is the 11" width of the paper.
   b. You should now have one tall, skinny cylinder and one short, fat cylinder.

Look at the two cylinders. Which cylinder do you think would hold more rice? Predict your answer.

c. Place the tall, skinny cylinder inside the short, fat cylinder and place them both into a cardboard box.

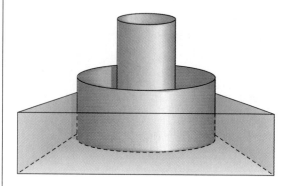

d. Fill the tall cylinder with rice. Now lift the tall cylinder out of the box so that the rice now fills the short cylinder. What happens? Record your results. Which cylinder holds the most? Why? Was your prediction correct?

▶ HOMEWORK PROJECT A

7. Find or make a right prism that has a triangular, hexagonal, or octagonal base. Sketch this prism. Measure and record the dimensions of this prism and find the volume.

8. Find and record the volume of a cylinder formed by an $8\frac{1}{2}$" x 11" piece of paper that has been folded so that the height of the cylinder is $8\frac{1}{2}$ inches. What is the radius of the base?

   a. Cut the paper as shown below.

   Tape the two halves end to end.

   b. Make a new cylinder and compute its height and volume. What is the radius of its base?

   c. Compare the radii, heights, and volumes of the two cylinders.

   d. Since the circumference of the new cylinder is twice as long as the circumference of the original cylinder, and the height is half as long, why aren't the volumes the same? Explain.

▶ GROUP PROJECT C

Find the surface area of a cylinder.

9. In Unit 4, you used nets to find the surface area of a cube. You can use a similar net to find the surface area of a cylinder.

   a. Look at the net below.

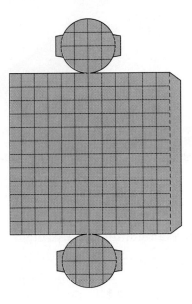

   Estimate and record the area of each circle. Then find and record the area of the rectangle.

   b. Discuss with your group how you could use these measures to find the surface area of the cylinder. Record your answer.

   c. Now count all of the units on the net. Record this answer. How does your previous answer compare to your net answer?

10. Use a paper towel tube and a piece of paper to make a cylinder.

   a. Trace around the end of the tube two times to make two circles, or bases. Cut out the bases.

b. Make one cut of the tube from end to end.

c. Open the tube and flatten the cylinder. What shape does it form? Use what you know about area to find the total surface area of the cylinder.

11. A circular cracker has a diameter of 4.7 centimeters. The thickness of each cracker is 0.5 centimeters. Thirty-four of these crackers are stacked on top of each other and packaged together with a sheet of plastic around them in airtight packs.

   a. Design the minimum size sheet of plastic you need for this pack. Consider how you would seal the ends of the package.

   b. Four airtight packs of crackers fill one rectangular box. Find the dimensions of this box.

   c. Find another rectangular box to hold the four airtight packs that has different dimensions. Which box has the least surface area?

▶**HOMEWORK PROJECT B**

12. Find a cylinder in your home or at school. Measure and record the surface area and volume of the cylinder. Show your work. Bring your cylinder and your data to class. Compare your cylinders with your classmates to see who has found the cylinders with the greatest surface area, the least surface area, the greatest volume, and the least volume.

13. Find or make two different-sized cylinders that have approximately the same volume.

WHAT'S THE PATH?

What's the path of a ball when you toss it up in the air and catch it? What's the path of Earth as it orbits around the sun each year? What's the path of a person riding on a playground swing? What's the shape of the curve used in a headlamp of a lighthouse reflector? What's the shape of the curve that describes a bathtub filling and emptying or a ride on a roller coaster? What's the shape of the curve that represents all possible dimensions for a rectangle of a fixed area? In this activity, you will investigate many different path patterns.

▶ GROUP PROJECT A

Find the number of different polygonal paths on a geoboard.

1.   **PORTFOLIO BUILDER**   A polygonal path is the union of a set of line segments with common endpoints. For the following investigation, use only those paths that move down or to the left. The pegs directly down or to the left of S can each be reached by only one path. The path traveled and the number of different paths to get there are shown below.

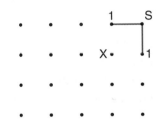

To go from S to point X, there are two different paths.

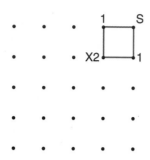

Since there are two different paths to reach X, we put a 2 next to X.

a. Use a geoboard or geoboard dot paper to find the number of paths from S to each of the other pegs on the board. Write the number of paths next to each peg or dot. As you go along, look for patterns in the numbers. On page 88 of your Portfolio Builder, list these patterns.

b. Now tilt your geoboard or geoboard paper so that point S is at the top center as shown below.

S
•

• •

• • •

• • • •

Complete the chart on page 88 of your Portfolio Builder to show the number of paths.

c. Next to each row in your chart, write the sum of the numbers in the row.

d. What patterns do you notice in the sums of the numbers in the rows? In the tenth row, what is the sum? In the nth row, what is the sum? The path pattern is known as **Pascal's Triangle** and was named after the French philosopher and mathematician Blaise Pascal (1623-1662).

▶ **HOMEWORK PROJECT**

2.   PORTFOLIO BUILDER   If you have a circular geoboard, use it to conduct this investigation. Otherwise, use circular dot paper.

a. **Inscribe** (place all vertices on the circle) different kinds of polygons (3, 4, 5, 6, 7, 8 or more sides). Show all diagonals. Into how many parts is the interior of the circle cut?

b. Count and record the number of parts on the chart on page 89 of your Portfolio Builder. Be sure to check various examples of each kind of polygon to be sure that your data is correct. Find a pattern in the data to let you predict for other cases. Test your prediction.

▶ GROUP PROJECT B

Use paper folding to make envelopes of curves and investigate their properties.

3. Use waxed paper to complete the following.

◆ Start with a sheet about 30 centimeters long. Make a line by folding the sheet near one end so that it is perpendicular to the sides of the sheet. Label this line *d*. Mark a dot labeled point F about midway between the sides of the sheet so that it is one of these distances away from line *d*: 2 cm, 4 cm, 6 cm, 8 cm, 10 cm, or 12 cm. Recall from Activity 1 that the distance is always the perpendicular distance. Each member of your group should pick one of these places for point F.

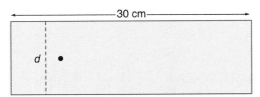

◆ Pick a point on line *d*. Put that point on top of point F and crease the paper. Pick another point on line *d*, place it on top of point F, and crease the paper. Repeat this 30 or 40 times so that you produce many segments.

a. Study your paper folding. What you should see with the many line segments is the shape, or outline, of a special curve. This outline is called an **envelope**, and the apparent curve is called a **parabola**. Trace along this curve with a dark marker.

b. To investigate your parabola, take some measurements. Mark and label eight to ten points along the curve. For each point, carefully measure two distances: first, to the point F, then to the line *d*. What appears to be true for every point of this curve?

c. Compare your curve to others in your group. Take all the sheets from your group and put one sheet over the other, so that all of your lines (*d*) coincide. How do the shapes of the curves vary? How are they alike? What is the effect of the placement of point F relative to line *d*? Did other group members find a similar result in Exercise 3b above? Share your results with the class. — *widens or narrows parabola*

d. COMPUTER OR GRAPHING CALCULATOR ACTIVITY If you have computer software for graphing or a graphing calculator, use it to make graphs of the three equations below.

  i. $y = x^2$
  ii. $y = 0.5x^2$
  iii. $y = 0.1x^2$

How do these compare with the envelopes of parabolas made in your group?

4. You can fold the envelope of another familiar, beautiful curve. Use the following steps.

   • Again, start with a sheet of waxed paper about 30 cm long. Using a plate or a bowl face down, trace a circle so it is clearly marked in the waxed paper. Select and label a point F in the *interior* of the circle. In your group, vary the placement of point F using these distances from the circle: 1 cm, 2 cm, 3 cm, 4 cm, 5 cm, or 6 cm.

   • Pick some point of the circle, fold it onto point F, and crease. Do this for 30 to 40 different points of the circle to make many line segments.

   a. Study the apparent curve made by the envelope. The curve is called an **ellipse**. Trace along this ellipse with a dark marker.

   b. To investigate your ellipse, you will again take some measurements. First, carefully fold the curve onto itself to find and mark the location of the reflection image of point F. Call it point G. Next, mark eight to ten points of the ellipse. For each marked point of the ellipse, measure

and record two distances: to the point F, and to the point G. Look for a pattern. What appears to be the same for all points of the ellipse?

   c. Compare your ellipse with those of other students in your group. As with the parabola, put each sheet over the other, so that the circles coincide. How do the shapes of the ellipses vary? How are they alike? What is the effect of the placement of point F relative to the circle? Did other group members find a similar result in Exercise 4b? Share your results with the class.

   sum of distance is constant

   d. COMPUTER OR GRAPHING CALCULATOR ACTIVITY If you have computer software for graphing or a graphing calculator, use it to make graphs of the equations below.

   not a function   *ellipse*

   i. $\dfrac{x^2}{2} + \dfrac{y^2}{10} = 5$

   ii. $\dfrac{x^2}{8} + \dfrac{y^2}{3} = 9$

   How do these compare with the envelopes made in your group?

5. Another famous curve can be folded by using a circle and a point. This time, choose point F outside the circle. As before, vary the placement of point F using these distances from the circle: 1 cm, 2 cm, 3 cm, 4 cm, 5 cm, or 6 cm.

   • The key step is the same as for folding the ellipse. Pick some point of the circle, fold it onto point F, and crease. Do this for 30 to 40 different points of the circle to make many line segments.

a. Study the apparent curve made by the envelope. It is called a **hyperbola**. Trace along the hyperbola with a dark marker. How does it appear to be similar to a parabola? How is it different?

b. To investigate your hyperbola, you will again take some measurements. As with the ellipse, carefully fold one curve onto the other curve to find and mark the location of the reflection image of point F. Call it point G. Next, mark eight to ten points of the hyperbola. For each marked point of the hyperbola, measure and record two distances: to the point F, and to the point G. Look for a pattern. What appears to be the same for all points of the hyperbola?

difference distances to constant

c. Compare your hyperbola with those of other students in your group. As with the ellipse, put each sheet over the other so that the circles coincide. How do the shapes of the hyperbola vary? How are they alike? What is the effect of the placement of point F relative to the circle? Did others find a similar result in Exercise 5b above? Share your results with the class.

d. COMPUTER OR GRAPHING CALCULATOR ACTIVITY If you have computer software for graphing or a graphing calculator, use it to make graphs of the equations below.

i. $y = \dfrac{12}{x}$

ii. $y = \dfrac{48}{x}$

How do these compare with the envelopes made in your group?

6. Many situations in the real world can be modeled with the parabola, the ellipse, or the hyperbola. Several applications were listed in the introduction to this activity. For each example below, which curve might be used?

♦ path of a ball when you toss it up in the air and catch it

♦ path of Earth as it orbits around the sun each year

♦ path of a person riding on a playground swing

♦ shape of the curve used in a headlamp of a lighthouse reflector

♦ shape of the curve that describes a bathtub filling and emptying

♦ the shape of a curve on a roller coaster ride

CHECKING YOUR PROGRESS

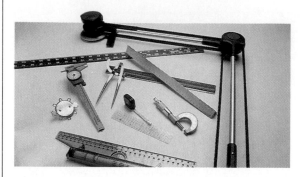

It is time to assess your learning of the mathematical ideas of this unit. You will complete an Individual Assessment and Partner Assessment related to the topics included in this unit. Then, your Unit Problem group will complete their cooperative work. You will prepare your individual report and need to be prepared to discuss it with the class.

PART I INDIVIDUAL ASSESSMENT

1. The word *geometry* comes from the Greek word geometria meaning Earth measurement. Select one of the following topics to write an essay on Earth measurement. Be sure to include the mathematics you have learned in this unit.

   ◆ What on Earth is measured and why?

   ◆ Look through magazines and newspapers and find articles and/ or pictures of geometry in life and in nature. Cut out or copy these items and describe the geometry in each.

2. Look at the plane figures that are not necessarily drawn to scale. What else might be wrong with the information in these figures? Write a paragraph for each figure, explaining your findings.

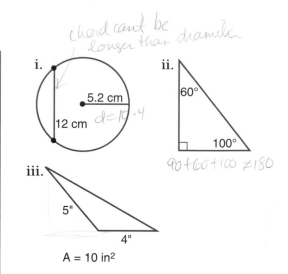

i. *chord can't be longer than diameter*

   5.2 cm
   12 cm   d = 10.4

ii. 60°   100°   *90 + 60 + 100 ≠ 180*

iii. 5"   4"   A = 10 in²

3. Find the area of each shape below. Show your work.

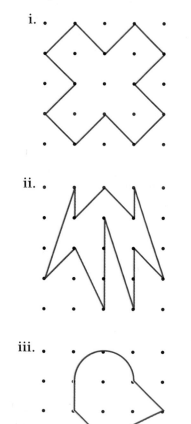

i.

ii.

iii.

PART II PARTNER ASSESSMENT

Use what you have learned in this unit to solve the following problem.

4. Imagine that Marcus decided to build a rectangular pen next to the garage. He could use the wall of the garage as one side of the pen, so he will need the fencing for the other three sides of the pen. Now, what are the dimensions that will give a maximum area for 48 feet of fencing? 64 feet? 100 feet? For any length, L?

PART III UNIT PROBLEM

Complete your cooperative and individual work. Write and present your report.

5. Meet in your group to complete your cooperative development of your project for the Unit Problem. Part of your grade will reflect what you contribute to the group activity. Listen carefully, ask questions, and make positive contributions.

6. As you develop your individual work on your project, you should consult with your teacher about your progress. As needed, cooperate with other members of your group to help each other. Part of your grade will reflect how thoroughly you use resources and develop your plans.

7. Submit your written report to be evaluated. Be prepared to revise your report based upon the feedback you receive. The major part of your grade will reflect the quality of your effort and your written report. An essential element will be the way you use key geometric ideas from this unit to communicate the results of this project.

8. Present an oral report of your project to the class. Prepare this presentation to fit in the time allowed. Think about how you can use the graphical displays you have prepared in your written report. Make it interesting and informative. Part of your grade will come from this oral report.

ENDING UNIT 6

PORTFOLIO BUILDER

Now that you have completed Unit 6, refer to page 90 of your Portfolio Builder. You will continue building your personal portfolio from items you made in this unit.

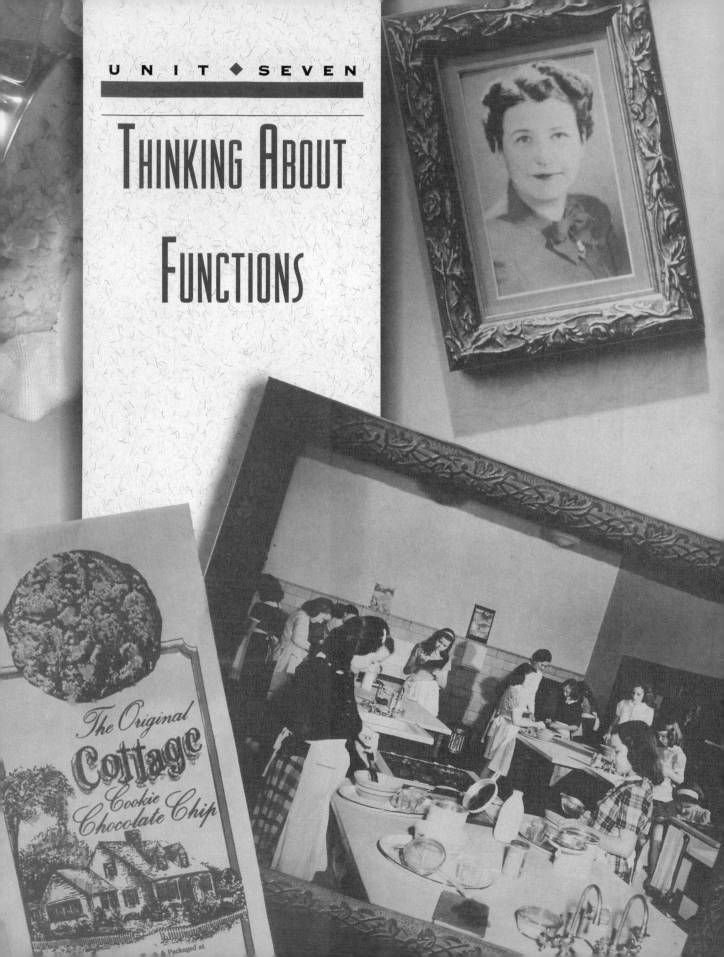

UNIT ◆ SEVEN

THINKING ABOUT FUNCTIONS

Can you imagine a time when there were no cars or television? How about chocolate chip cookies? Believe it or not, chocolate chip cookies were "invented" in 1933 by Ruth Graves Wakefield.

Mrs. Wakefield learned to love cooking from her grandmother. After graduating from Massachusetts State Teachers College in 1924, she taught home economics. But she and her husband dreamed of opening a small inn. So with more courage than money, they purchased an old toll house that had been the halfway point between Boston and New Bedford.

According to legend, Mrs. Wakefield was in a hurry baking a batch of cookies. Instead of melting the chocolate, she broke the chocolate bar into small pieces and added them to the batter - and "chocolate crispies" were born.

Meanwhile, the Nestle company was thinking of discontinuing the semi-sweet chocolate bar. But sales began to skyrocket around the Boston area. Representatives of Nestle were excited to learn about the popular new cookie, and in 1940, Nestle bought the name "Toll House" from the Wakefields. Shortly, Nestle began making chocolate chips, just for these cookies. To this day, Nestle still prints Ruth Wakefield's recipe for Toll House Cookies on the back of their chocolate chip bags.

ABOUT THIS UNIT

Throughout this course, you have investigated many different mathematical situations. You have often been asked to gather and organize data so that you could detect patterns to answer questions or reach conclusions. By looking at patterns, you may have found relationships that could be stated in formulas or rules.

One very important way to represent data is with graphs. Graphs can be used to show how numbers vary, and by interpreting graphs, we can find relationships that may lead to an expression with variables.

In this unit, you will explore variables, expressions, equations, and graphs in order to grasp more clearly some of the basic ideas of algebra. You will learn to make graphs from equations and to interpret graphs in order to find an equation. You will learn to solve equations graphically and algebraically. These experiences should help you build a solid background in order to study algebra and other higher mathematics with greater success.

▶ UNIT PROBLEMS

Ideas for the four Unit Problems are presented below. Early in your study of this unit you should select one of the topics, **housing**, **wealth**, **health**, or **population**, to develop your Unit Problem. By the end of this unit, you should use the mathematics you have learned to help complete your Unit Problem.

1. HOUSING

The construction of a building involves many ideas from mathematics. Equations and graphs are used by architects and construction engineers to help determine methods and costs of building.

a. Lumber is a basic building material. Find out what a board foot of lumber means. Develop equations, charts, or graphs to help compute board feet. Milled lumber is cut from logs. Find out how boards are cut from a log, and develop a formula or chart for computing the number of board feet in a given log of diameter d and length l. Find other equations used in producing or using lumber. For example, find or develop an equation to tell how much lumber per square foot of floor is needed to build an average house.

b. Concrete is another important and widely-used building material. Concrete is ordered and sold by the cubic yard. Develop an equation to find the amount of concrete needed for a foundation wall, garage floor, or sidewalk. Study how concrete is mixed and how the strength can vary. Find what formulas are used to guide the mixing of concrete. Study the costs of using concrete and develop a formula for finding costs. Complete a cost analysis for the concrete in a building you select.

c. Many houses and offices are air conditioned. Investigate how engineers determine the appropriate air conditioning system for a building. What formulas are used? Apply what you learn to determine an air conditioning plan for a structure, such as your house or school. Use equations and graphs to report your results.

d. Investigate the use of a solar heating system in your home. Develop a formula for determining how long it would take the savings in utility bills to pay for the installation. Also include any savings due to federal tax credits.

2. WEALTH

The financial pages of the daily newspaper contain many numbers. Graphs and equations can help to make sense of all of this data.

a. Study the information on the financial pages. Select some section and investigate how the numbers behave. For example, choose five stocks and study how their prices vary over at least 10 days. Make a graph. Fit a line to the data and write its equation. Use it to predict the prices of the stock in a week, then check your predicted stock prices against the actual price in a week.

b. Study the world currency market. How do various foreign currencies change in value compared to the U.S. dollar? Try to use data from many months or years. Make graphs and write equations to demonstrate any

trends. Develop formulas for converting between currencies. Study and explain how the value of some currency changes compared to the price of gold.

c. Investigate the commodities market, using the financial pages. What are commodities, and what affects their value? Pick some commodities and study how their value changes. Use graphs and equations to describe how they vary. What stocks might have their value tied to the value of certain commodities? Study possible connections.

3. HEALTH

To be healthy, we all need drinking water and good food.

a. One of the most crucial aspects of our health is safe drinking water. Study the problem of drinking water and sanitation around the world. Find data for which you can write equations and make graphs to show what is happening. The problems of ground water are obviously linked to available drinking water. What is happening to our ground water, and why? What is happening with freshwater pollution, such as acid rain?

b. Millions of people are hungry. Study the problem of hunger, starvation, and other food-related problems. Develop equations to show rates of change in the problems. What are the trends? Study related problems, such as the decrease in grazing land, the loss of topsoil, the decline in the number of farmers, or the increase in population. Find out about the major sources of protein, how we Americans get protein, and what our protein costs to produce compared to other countries. Use graphs and charts to present your information.

4. POPULATION

Many factors affect the size and change of a population.

a. Study population growth data for countries around the world. Which countries have the fastest growth rates? Are there countries where the population is declining? Use graphs to show their growth or decline over the past 50 years. Write or find equations that summarize the growth change. Use them to predict the sizes of various countries in the future. Study the world population change, and write an equation to predict future growth.

b. Investigate the birth rate and death rate for different countries. How have these rates changed in the past 50 years? Write equations to show each rate, and graph them on the same axes for comparisons.

c. Select five countries, and study the principal causes of death. How have the rates of death by each cause changed in the past 50 years? Write equations and make graphs to show what you find. Compare the results across the five countries. What can you conclude?

IN SEARCH OF THE UNKNOWN

In this activity, you will explore more thoroughly how variables behave. You will use patterns, tables of values, and graphs to help you investigate the possible relationships.

▶ PARTNER PROJECT A

Use pattern blocks to find and extend patterns.

1. **PORTFOLIO BUILDER** Use pattern blocks to help you extend each pattern below. Then complete each chart on pages 91 and 92 of your Portfolio Builder.

i.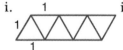

ii.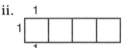

iii.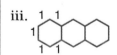

iv.

a. For each pattern, explain in words what happens to the perimeter as you add a new block.

b. Extend your charts to include the **nth term.**

c. Suppose a shape has 25 sides and is lined up with a common side like the shapes above. Make a chart on page 93 of your Portfolio Builder to show the number of 25-sided shapes and the perimeter of each. Extend this chart to include the nth term.

d. On graph paper, make a graph of the results from the pattern charts on the same axes. Let the horizontal axis represent the number of polygons, and the vertical axis represent the perimeter. What conclusions can you make about the points you have plotted?

e. Look back at the charts you made. In this situation, the perimeter is the **dependent variable**. What does the perimeter depend on in each case?

2. 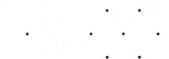 Look at the patterns of figures below.

i.

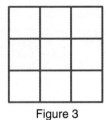

Figure 1   Figure 2   Figure 3

ii.

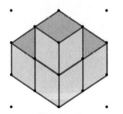

Figure 1

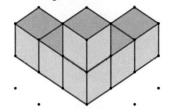

Figure 2

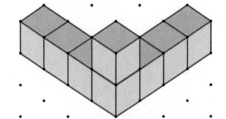

Figure 3

iii.

Figure 1   Figure 2

Figure 3

a. In each of the patterns, build and draw the next figure on pages 94 and 95 of your Portfolio Builder.

b. List these terms and the next four terms of each in the chart on pages 94 and 95 of your Portfolio Builder.

c. In each, describe the patterns that you notice.

d. Continue each chart to include the *n*th term.

e. In the first pattern, the number of squares depends on which term it is. The term number is the **independent variable** because it does not depend on anything. Make a graph of the results in each chart. Place the independent variable along the horizontal, or *x*-axis, and the dependent variable along the vertical, or *y*-axis.

▶ INDIVIDUAL PROJECT

Use patterns to solve problems.

3. The students at Yerba Buena High School want to have a campus beautification project. In their quad, the student council wants to build 100 flower beds and surround them with triangular stepping stones according to the pattern shown below.

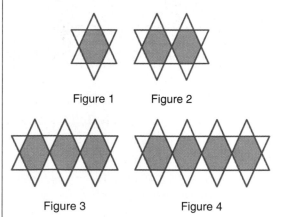

Figure 1   Figure 2

Figure 3   Figure 4

   a. Organize your work using charts and diagrams. Look for patterns to find a general rule. Explain why your rule works.

   b. How many stepping stones will the student council need?

   c. Predict the number of stepping stones needed for *n* flower beds.

   d. Explain the pattern. Make a graph of this pattern. What are the independent and dependent variables? Explain. Save this write-up for your portfolio.

▶ PARTNER PROJECT B

Use string and a pair of scissors to help you solve the following problems.

If you cut a piece of string a number of times, can you figure out the number of pieces of string you will have?

cuts	1	2	3	n
pieces	2	3	4	n+1

4. Cut a piece of string one time.

   a. How many pieces of string do you have?

   b. Now make a second cut in the original piece of string. How many pieces of string do you have now?

   c. On page 96 of your Portfolio Builder, record the number of cuts and the number of pieces you have made. Continue your chart to include 3, 4, and 5 cuts.

d. Explain the patterns you see.

e. Write a rule to determine how many pieces there will be if there are *n* cuts in your string.

5. Fold the string as shown below. How many pieces will be formed by the cut?

c	1	2	n
P	3	5	2n+1

a. 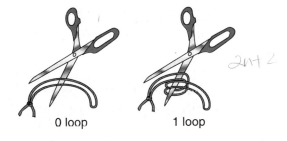 Complete the chart of page 96 of your Portfolio Builder to show the number of cuts and the number of pieces.

b. Look at the string as you cut each time. Explain the pattern you see.

c. Write a rule to determine how many pieces of string there will be if there are *n* cuts in your string.

6. Loop a piece of string around your scissors as shown below.

n+2 from n loops

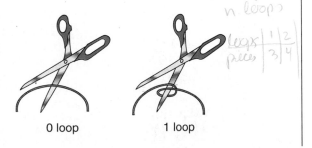

loops	1	2
pieces	3	4

0 loop   1 loop

a. This time, consider how the number of loops affects the number of pieces formed after only one cut. Complete the chart on page 96 of your Portfolio Builder to show the number of loops and the number of pieces.

b. Look at the string as you cut each time. Explain the pattern you see.

c. Write a rule to determine the connection between *n* loops and the number of pieces.

7. Now tie the ends of the string together before you loop the string around the scissors.

2n+2

0 loop   1 loop

a. How many pieces of string are there for each number of loops? Record the information on page 96 of your Portfolio Builder.

b. Look at the string as you cut each time. Explain the pattern you see.

c. Write a rule to determine the connection between *n* loops and the number of pieces.

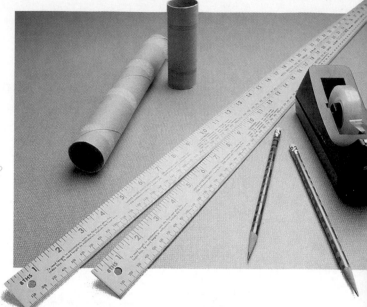

8. How many pieces of string will you have if the string is folded in each way shown below before being looped around the scissors and cut one time?

a.  $2n+3$

b.  $3n+4$

c.  $4n+5$

 On page 96 of your Portfolio Builder, record the information in the chart. Look for patterns. Then write a rule to predict the number of pieces for n loops.

9. Share your rules from Exercises 4 through 8 with your class. Then identify the dependent and independent variables in each of those exercises.

10. Graph each group of points listed below on separate graphs.

   a. Exercises 4 and 5

   b. Exercises 4 and 6

   c. Exercises 5, 7, and 8a

   d. Exercises 6 and 7

   e. Exercises 8a, 8b, and 8c

   f. Look at the graphs you made. Explain the similarities and differences you see in the graphs.

   g. If you connect the consecutive points with line segments and extend them, what geometric curve is formed?

▶ **PARTNER PROJECT C**

Perform an experiment and analyze data.

11. Perform the following investigation.

   ◆ Use the cardboard tube from a roll of bathroom tissue or paper towels.

   ◆ Measure the length and diameter of the tube. Record this information.

   ◆ Tape two meter sticks, vertically and end-to-end, to the wall or chalkboard.

   ◆ Along the floor, measure and mark 100 cm, 200 cm, 300 cm, and 400 cm from the meter sticks.

◆ Choose one person to be the viewer. This person needs to remain the same throughout this investigation.

◆ Have the viewer stand with their instep on one of the measured marks. Record this distance.

◆ Aim the tube at the meter sticks. Have the partner stand by the meter sticks and use two pencils or two fingers to determine the amount of the meter stick that can be seen by the viewer. When the viewer says that he or she can barely see the two pencils, measure the distance between the pencils and record this measure as the field of view.

◆ Repeat this activity from each of the marked distances.

a. Record the distances and the lengths of your field of view in a table. **b.** What is the independent variable? How do you know? What is the dependent variable? How do you know?

c. Graph the data. On which axis did you plot the dependent variable? Why?

d. Draw a line to connect the points. Extend the line in both directions. If you can see 80 cm of the meter stick, how far away are you? Use your graph to help you. If you were 18 m away, what would your field of view be?

e. Find the ratio of the diameter of the tube to the length of the tube. Then find the ratio of the length of the field of view to the distance used. Compare the ratios. What conclusions can you make?

f. Write a rule or formula for this investigation. Be sure to identify the variables you use.

g. Perform the investigation again using different-sized tubes. Graph the data. Then write a new rule or formula. How does it change? Why? Compare these lines with the line you graphed in Exercise 11d. How are they the same? How are they different? Explain.

▶ HOMEWORK PROJECT

12. Copy the patterns below.

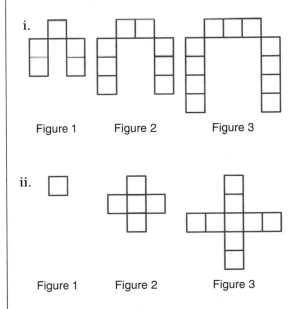

i.

Figure 1   Figure 2   Figure 3

ii.

Figure 1   Figure 2   Figure 3

a. Draw the next two figures for each pattern.

b. Make a chart to compare the figure number to the number of squares.

c. Explain the patterns you see. Write a rule for extending each pattern to the nth term. Check to see that your rule works for $n = 3$. Then use your rule to find the 63rd term.

13. Copy the pattern below.

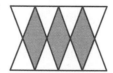

Figure 1   Figure 2   Figure 3

a. Draw the next two figures for each pattern.
b. Make a chart to compare the number of rhombi to the number of small triangles.
c. Explain the patterns you see. Write a rule to extend the chart to include the nth term. Check to see that your rule works for $n = 3$. Then use your rule to find the 50th term.

14. Julian looked through two different view tubes. The graphs of the curves for these tubes are shown below.

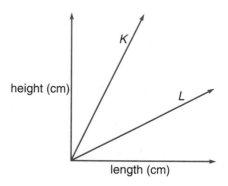

a. How are the dimensions of tube K different from those of tube L?
b. Does the steepness of the line depend on the width of the tube, the length of the tube, or both? Explain.
c. If the length of the tube remains the same and the width of the tube varies, what happens to the steepness of the line? Explain.
d. If the width of the tube remains the same and the length of the tube varies, what happens to the steepness of the line? Explain. Be prepared to discuss your reasoning in class.

COMPARING GRAPHS OF LINES

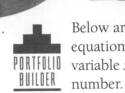

The idea of families is something all of us are familiar with. We know that human families share common characteristics. In science, we learn about families in the animal world as well as in the plant world. Think about some families in each of these settings and name some common characteristics.

In this activity, we are going to look at equations in mathematical families and their graphs and discuss their common characteristics.

▶ **GROUP PROJECT**

Analyze families of lines.

It is recommended that the following exercises be done with a graphing calculator or computer.

1. **PORTFOLIO BUILDER** Below are some families of equations that use the variable x, where x is any real number.

Family 1

$y = x$

$y = 2x$

$y = 3x$

$y = 4x$

$y = 5x$

Family 2	**Family 3**
$y = x + 1$	$y = x - 1$
$y = x + 2$	$y = x - 2$
$y = x + 3$	$y = x - 3$
$y = x + 4$	$y = x - 4$
$y = x + 5$	$y = x - 5$

Family 4	**Family 5**
$y = 2x + 1$	$y = 2x - 1$
$y = 2x + 2$	$y = 2x - 2$
$y = 2x + 3$	$y = 2x - 3$
$y = 2x + 4$	$y = 2x - 4$
$y = 2x + 5$	$y = 2x - 5$

As you go along in this exercise, keep a record on pages 97 through 101 of your Portfolio Builder of sketches, charts, ideas, and answers to questions. Keep these items for your portfolio.

    a. In your group, study each family. How are the equations within each family alike? How are they different?

    b. Graph each family on a computer or graphing calculator and compare the graphs you see. How are they alike? How do they differ?

    c. For each equation within each family, make a table to find the coordinates of three points. To understand the relationship, study how the value of y varies as the value of x changes.

    d. Graph each family of equations on the same set of axes so that you can compare them more easily. Label each graph with its equation. Discuss your graphs in your group.

    e. Now compare the families. How are they alike? How do they differ? How does a change in the equation between families seem to affect the graphs? Write your explanations on page 102 of your Portfolio Builder.

    f. Write a paragraph of your conclusions about these families on page 102 of your Portfolio Builder.

2. COMPUTER ACTIVITY Use a spreadsheet program to make the table of values for each of the equations in Exercise 1.

3. COMPUTER ACTIVITY Complete the BASIC program below to make a table of values for the first family in Exercise 1.

10 REM Comparing Algebraic Expressions

20 PRINT "x 2x ___ ___ ___"

30 FOR x = 1 TO 20

40 PRINT x; ___; ___; ___; ___

50 NEXT x

60 END

RUN the program and use the results. What does the program choose x to be? Change the program to find the tables for each of the other families.

4. In Unit 6, you investigated properties of polygons. Use the results recorded on page 82 of your Portfolio Builder.

    a. In Unit 6, Activity 2, Exercise 8, you studied the areas of families of triangles. Recall the formula you used to find the area of a triangle. In the examples suggested, the triangles all had the same base and had heights that were the same measure. What happens if you work with the same base, but let the height, h, vary? Investigate this situation. Make a chart and a graph of the ordered pairs, (height, area) when the base is 4 units. What is the shape of the graph? Write an equation ($y = ?$). Make graphs on the same axes when the base is 2, 6, 8, and 10 units and the height varies. Compare the

graphs. Write an equation for each. What can you conclude?

b. Again, study the areas of families of triangles. This time, fix the height and let the measure of the base vary. Make a chart and graph of the ordered pairs, (base, area) when the height is 2 units. Make graphs on the same axes when the height is 4, 6, 8, and 10 units. Compare the graphs. Write an equation for each. What can you conclude?

▶ HOMEWORK PROJECT

5. Given the equation $y = 5x + 3$, write four other equations that fall in the same family. Then graph this family.

6. Given the equation $y = -3x + 4$, write four other equations that fall in the same family. Then graph this family.

FROM WORDS TO EQUATIONS

Research scientists, engineers, and economists collect data and study the relationship that pieces of data have to each other to make predictions or theories about trends. Scientists graph their data to see if there is a relationship. If a relationship exists, they see if it can be used as a model to develop a formula or rule.

Additional information can be found from a graph by extending the graph and **extrapolating**, or estimating values beyond those that are known, or by **interpolating**, or estimating values of a variable between other values that are known.

In this activity, you will look for relationships, write equations, and analyze results.

▶ **INDIVIDUAL PROJECT**

Use variables and expressions to solve problems.

1. Systolic blood pressure is the contraction of the heart that forces blood onward. A formula for systolic blood pressure in a human is $\frac{1}{2}$ of the age plus 100 or $s = \frac{1}{2}a + 100$.

   a. Find s for your age and several other people who are younger and older than you.

   b. Identify the independent and dependent variables. On a set of axes, plot some of the points and make a graph. Is this a line? Why or why not?

   c. Estimate the answers for the following questions by looking at the graph you made.

   ◆ Estimate what the systolic blood pressure would be for a person who is 50 years old.

   ◆ How old would you estimate a person to be if his or her systolic blood pressure was 121?

   d. If you have access to a blood pressure cuff and information about using it, compare your actual systolic blood pressure to the estimate you got with the formula.

2. Think about the ages of your family members. You can use algebraic equations to explore age relationships in your family.

   a. How old was your mother when you were born? Beginning with your birth and up to your current age,

make a chart to show the ordered pairs, (your age, your mother's age). Graph these ordered pairs. What is the shape of the graph? Write an equation to find your mother's age, given any number x for your age. What can you conclude?

b. On the same axes as above, make a graph for your age and each of your family members. Compare the graphs. Write equations to find each person's age, given any number x for your age. What can you conclude?

c. What is your age when your mother is twice your age? How many times will this happen? Why? Write an equation for this situation. Study the equation. What can you conclude?

d. Find your age when each family member is twice your age, or you are twice the age of a younger sibling.

Write equations for each one. Be sure to define the variables. What can you conclude?

e. Investigate your age when your mother is three, four, and five times your age. Write equations for each case. Discuss your results.

▶ PARTNER PROJECT

Use variables and expressions to solve problems.

let x = time
y = miles

3. The speed of light is about 186 000 miles per second. Choose two different variables, define them, and write an equation to determine how far light travels in a given time.

a. How far will light travel in one year? 5 years? 10 years?

b. Which is the independent variable? Which is the dependent variable? Explain.

c. Graph your ordered pairs. Carefully plan the scale of your axes.

d. Use your graph to estimate the answers to the following questions.

 ◆ The sun is approximately 93 000 000 miles from Earth. How long does it take light from the sun to reach Earth?

 ◆ Light from the star Sirius takes about 8.8 years to reach Earth. About how far in miles is Sirius from Earth?

4. One of the most important means of transportation today are commercial airlines. Winds affect the speed of an airplane. Once an airplane takes off, it is

carried along by the mass of air that surrounds it. If the air is still, the ground speed will equal the air speed.

a. If, however, the wind is coming toward the tail of the plane, called a **tailwind**, the airplane's ground speed as measured from the ground is found by adding the air speed to the speed of the wind. Study the effect of a tailwind for an air speed of 400 mph. Make a chart and graph the points. What is the shape of the graph? Write an equation to find the ground speed, given any tailwind speed, t. Think about a tailwind speed of 40 miles per hour. Locate this point on your graph. What is the ground speed?

b. If, however, the wind is coming toward the front of the plane, it is called a **headwind**. A headwind slows an airplane, so the ground speed is the difference of the air speed and the headwind. Study the effect of a headwind for an air speed of 400 mph. Write an equation to find the ground speed, given any headwind speed, h. Make a graph of the pairs. Find the ground speed for a headwind speed of 30 miles per hour.

c. Compare your results for tailwinds and headwinds. How are the equations alike? How are they different? How are the graphs alike? How are they different? Write a conclusion about your results.

▶ HOMEWORK PROJECT

5. Study the following statements.
- ◆ The sum of two numbers is 5.
- ◆ Three times one number added to another number is 18.
- ◆ The difference between two numbers is 4.
- ◆ One number minus two times another number is 5.
- ◆ The first number of two numbers is always 2.
- ◆ The second number of two numbers is always 0.

a. For each of the statements above, find at least three ordered pairs for each statement. Graph these ordered pairs. Extend the graph and list two more ordered pairs that lie on this graph.

b. Write an equation for each statement. Use one variable for the first number and another variable for the second number. Which is the independent variable? Which is the dependent variable? Explain. Label each graph with its appropriate equation.

Functions And Their Graphs

One of the most powerful ideas in mathematics is the concept of a *function*. The basic idea of a function is that two quantities are related in some way. The value of one quantity may depend on the value of another quantity. For example, the amount of medicine that a person should take depends on the weight of that person. Changing the weight changes the medicine dose, but changing the medicine dose does not make the person weigh more or less.

A function can be represented in many ways. It can be described as a rule or equation that makes clear how pairs of numbers are related. It can be described as a list or a set of the ordered pairs. It can be described as a graph of the ordered pairs. You have been exploring situations throughout this course, some of which are functions.

In this activity, you will see more clearly how we can think about functions. You will investigate various situations to see which relations are functions.

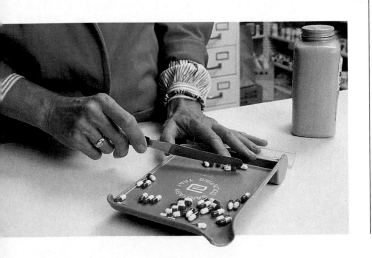

▶ GROUP PROJECT

Investigate the idea of a function.

1. There are many real-world situations in which the value of one variable depends on the value of another variable. A **function** is a set of ordered pairs that has, at most, one value for the dependent variable for each value of the independent variable. For example, in treating fevers in children, the following dosage chart appears on the back of an aspirin bottle.

DOSAGE CHART

WEIGHT (POUNDS)	NUMBER OF CAPLETS
under 48	consult physician
48 – 59	2
60 – 71	$2\frac{1}{2}$
72 – 95	3

  a. The number of caplets you should give a child depends on the weight of that child. If the chart also had a column that said that a child weighing between 65 and 68 pounds should be given 6 caplets, what might happen?
  b. The chart on the back of an aspirin bottle has to be a function. Why?
  c. List some ordered pairs for this example.

d. As a class, discuss other similar situations .

2. The following table shows one way of determining the maximum heartbeats per minute in terms of your age.

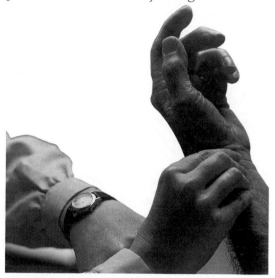

HEART BEATS

AGE (a)	NUMBER OF BEATS PER MINUTE (b)
10	164
20	156
30	148
40	140

The formula is $b = 172 - 0.8a$ where b is the number of beats per minute and a is the age.

a. Use the formula to find the maximum number of beats per minute for your age and the ages of two other people.

b. Make a graph to show the ordered pairs. What is the dependent variable? Why? What is the independent variable? Why? Is this a function? Explain why or why not.

c. What other exercises in this course have you done that are functions? List two of them and explain why they are functions.

3. In your group, investigate the height, weight, and age of people from the data table below.

AGE, HEIGHT, WEIGHT

NAME	AGE	HEIGHT (IN.)	WEIGHT (LB)
Joe	25	69	180
Anh	18	62	105
Julie	24	62	105
Josephine	80	56	95
Yolanda	16	66	130
Ray	53	69	195
Javier	16	72	195
Karla	50	64	120
Keisha	13	57	78
Sam	14	62	200
Kendra	3	27	30
Tran	8	50	85
Tomás	52	69	168

a. Make three different graphs to show the ordered pairs, (age, height), (age, weight), and (height, weight). On each graph, label the points, A, B, C, and so on to represent each person. Study the graphs. What relationships can you find?

b. Look at the graph of the ordered pairs for (age, height). Which people have the same age? Which person is the oldest? Which people have the same height? Which point represents the tallest person? The person who is tallest and oldest? Youngest and shortest? Are there people that have the same age, but different heights? Look back at the definition of a function. Explain why this set of ordered pairs is not a function. How could you record the age data differently and change the scale on the age-axis to possibly make this set of pairs a function?

c. Look at the graph of the ordered pairs (age, weight). Find the points for each of the following.
   ◆ the youngest *(3, 30)*
   ◆ the "average" age and weight
   ◆ the oldest and heaviest *(53, 195)*
   ◆ the lightest *(3, 30)*
   ◆ the same weights but different ages *(18, 105)*
   Are there any cases with the same *(24, 105)*
   age, but different weights? If so, is *(53, 95)*
   this group of ordered pairs a *(16, 195)*
   function? Why or why not?

d. Study the graph for the ordered pairs, (height, weight). Write a paragraph describing your findings. What relationships did you find? Tell whether this set of ordered pairs is a function, and explain why or why not.

4. There are some interesting bone measures in the human body. Many of these measures yield ratios that are highly consistent, regardless of the size and shape of the human body. Paleontologists make predictions about ancient animals from studying skeletal measurements.

a. In your group, find, to the nearest millimeter, the following measures for each person.

circumference of:

- head
- neck
- wrist
- closed fist

length of:

- tibia (lower leg bone from ankle to kneecap)
- foot
- hand span (tip of thumb to tip of little finger with hand spread open) radius (lower arm bone from wrist to elbow)
- wing span (finger tip to finger tip with arms spread wide open)

- naval to floor
- height ⟶ *thigh bone.*

Tabulate your group's data and carefully check this data to make sure that it is reasonable.

b. In your group, prepare a graph for each of the following ordered pairs of measurements.

- (height, wing span)
- (height, radius)
- (closed fist, foot length)
- (wrist, hand span)
- (head, neck)
- (tibia, height)
- (height, naval to floor)
- (radius, foot length)

In addition, have each member make a graph of one other ordered pair. Study the graphs. What are the shapes? Which of the data sets are functions, and why?

c. On your data chart, calculate and record the ratios for each person, using the ordered pairs named in Exercise 4b above. Study the results. What patterns and relationships do you find? Do any ratios appear to be constant across different people? How does such a constant ratio show up when you graph the data? Explain.

d. Exchange this data with another group. Analyze their data, using graphs and forming ratios as suggested above. What conclusions can you draw? Discuss your observations with the other group. Save the data chart, graphs, and your conclusion to add to your portfolio.

e. Organize an investigation of these measurements for a sample of persons outside of school. Record the data, make graphs, compute ratios, and look for patterns. Write a report of your results. How did this sample compare with your group?

f. In addition to paleontology, find ways that these ideas about body ratios are used. Be prepared to share your findings with the class.

▶ PARTNER PROJECT

Making and interpreting graphs.

With a partner, select two of the following exercises to complete. Be ready to share your results with the class.

5. Study the following clues.

   ◆ A and B are the same age, but B is heavier.

   ◆ C and D weigh the same and are each lighter than B, each younger than A or B, and C is the youngest of these.

   ◆ E is heavier than C, but is the same age as C.

   ◆ F is the oldest and heaviest.

   a. Sketch a graph of the ordered pairs (weight, age) that satisfies the clues. Compare your graph with your class. How are they alike or different?

   b. From your graph, can you tell whether this set of ordered pairs is a function? Explain why or why not.

6. Recall when you were studying triangles that the sum of the interior angles of any triangle was 180 degrees. If we take any polygon and subdivide it into triangles from any one vertex, we can determine the number of degrees in the interior angles of the polygon. If the number of sides, n, equals 3, then the number of degrees of the interior angles is 180. If the number of sides, n, equals 4, then you have 2 triangles and the number of degrees is 2×180.

   a. Write the ordered pairs, (number of sides, number of degrees) for polygons with 5, 6, 7, and 10 sides. Graph the pairs.

   b. How would you determine the number of degrees for a polygon that has 40 sides?

   c. Is this a function? Explain why or why not.

7. Study these clues about the gas mileage for several different cars.

   ◆ Cars A and B drive the same distance, but B uses more gas. Car C uses the same amount of gas as does car D, but C travels the same distance as car A.

   ◆ Car D travels further than does car A.

   ◆ Cars E and F use the same amount of gas, but car F travels further than does car E, which travels the same distance as car D.

   a. Sketch a graph of the ordered pairs, (gallons of gas, miles) that satisfies the clues. Which car has the best mileage rate? The worst? Explain your reasoning.

   b. From the graph, can you tell whether this set of ordered pairs is a function? Explain why or why not.

8. It is important to conserve gasoline, since the world's supplies are limited. How does the gasoline use in a car vary with the speed driven?

a. Suppose you drive on a freeway where the minimum speed is 40 mph and the maximum speed is 65 mph. Assume you have collected the following data from trips on a freeway.

FREEWAY DATA

SPEED (MPH)	GASOLINE (GAL)	DISTANCE (MILES)
40	2.0	58.6
45	6.1	184.2
50	10.2	284.1
55	15.6	427.4
60	8.5	224.4
65	16.3	422.2
70	9.7	233.8

Compute the mileage (miles per gallon) for each speed. Graph the ordered pairs, (speed, mileage). What is the shape of the graph? Explain whether it is a function.

b. Did you question the data above? Do you think anyone could actually drive at exactly the given speeds for the reported distances? If not, what do you think is meant by the reported speed? How would it be found?

9. How does the mileage vary for different speeds for different kinds of vehicles? Investigate the average miles per gallon of different cars. Obtain data on new car specifications from car dealerships. Try to include as many of these as possible: subcompacts, compacts, mid-size, full-size, sports car, luxury car, motorcycle, small pick-up truck, off-road truck, full-size pick-up truck, or van. As in the chart above, collect data for various speeds. Graph the data for each vehicle, using the same axes. Are these functions? Why or why not? Write a conclusion of your findings.

▶ **HOMEWORK PROJECT**

10. Make up your own set of clues for six people that deal with various values for two variables. Try to think of interesting variables to compare. Then, in class, exchange clues with another person to sketch a graph and discuss the possible relationships for each other's situation. Decide if the set of points that satisfy their clues is a function.

11. Lisa wanted to conduct a survey to find out how much her local Burger King took in just from selling Whoppers. She stood by the counter one day at noon and recorded the number of Whoppers sold every 10 minutes.

a. When Lisa analyzed her data, she identified two variables that were changing. What were they?

b. Lisa made a table of ordered pairs for every 10 minutes, knowing that one Whopper™ cost $1.98. Make Lisa's table. Is it a function? Explain.

c. How many Whoppers™ were sold during the lunch hour? How much money was made?

d. If the restaurant had similar business for the next half hour, estimate the total amount of Whoppers™ sold and the total amount of money made in $1\frac{1}{2}$ hours.

WHOPPER™ SALES

TIME	NUMBER OF WHOPPERS™	MONEY TAKEN IN
12:00 – 12:10	40	$79.20
12:10 – 12:20	38	$75.24
12:20 – 12:30	45	$89.10
12:30 – 12:40	41	$81.18
12:40 – 12:50	46	$91.08
12:50 – 1:00	52	$102.96

EXPLORING LINEAR FUNCTIONS

You have seen many situations where the points of a graph lie on a line. The line is often used to model or describe relationships between two variables. You may already recognize equations whose graphs are lines.

In this activity, you will learn more about the properties of **linear functions**. You will see how you can use the form of an equation to tell something about the line and how the line can be used to find the equation.

▶ ## GROUP PROJECT A

Investigate properties of linear functions.

The parts of an algebraic equation written in the form $y = ax + b$ have names. The a stands for any number that is a multiplier of x; a is called the **coefficient** of x. b stands for any number that is added or subtracted from the dependent variable. It is called the **constant**.

1. In your group, review your results on pages 97 through 101 of your Portfolio Builder, where you studied families of equations.

PORTFOLIO BUILDER

  a. Look at the graphs you made for each family. What is the shape of each graph? Are these functions? Why or why not?

  b. The equation $y = ax$ could be used to describe all of the equations for family 1. What are the values for a in family 1? Write $y = ax$ beside family 1 on page 97 of your Portfolio Builder.

  c. The equation $y = x + b$ could be used to describe all of the equations for family 2. What are the values for b in family 2? Write $y = x + b$ on page 98 of your Portfolio Builder, next to family 2.

  d. What similar equations could be used to describe families 3, 4, and 5? Discuss these in your group. Write these equations on pages 99, 100, and 101 of your Portfolio Builder next to their family.

  e. Make graphs for a family 6 of the form $y = 3x + b$ and a family 7 of the form $y = 3x - b$ where b varies in each case.

  f. Study these graphs and your earlier graphs. What is the shape of the graphs? Are they functions? From all of the graphs, what can you conclude about how the value of b affects the graph? Test your theory by trying it with a new equation. Discuss your theory in your group. Write your theory on page 103 of your Portfolio Builder.

  g. Look across the families to see how a affects the graph. Write a theory and test your theory with a new value of a. Discuss your theory in your group. Write your theory on page 103 of your Portfolio Builder. What happens to the graph if a is equal to 0?

  h. Are horizontal lines functions? Are vertical lines functions? What is the equation of any horizontal line? What is the equation of any vertical line?

▶ **PARTNER PROJECT**

Use properties of linear functions to solve problems.

Solve at least two of the following problems. Present your results to the class.

2. Suppose your club decided to raise money by sponsoring a concert.

   a. If the band and other expenses cost $2 000, and you decide to charge $10 per ticket, then the profit, p, from t tickets sold would be $p = 10t - 2\ 000$. Study this equation. Explain why you know it is a linear equation. Use any theories about linear equations you found in Exercise 1. What can you tell about its graph by simply looking at the equation? Make a chart and graph the equation to check your predictions.

   b. Pose questions about this situation. For example, what is the value of p when you "break even?" How many tickets would you have to sell to "break even?" What is the significance of this point on the graph? What if you don't sell any tickets? Where is this point on the graph, and what is its significance? What can you say about the steepness of the line, and how is this indicated by the equation? Pose other questions, and be ready to share these with the class.

   c. Suppose someone in your club suggests that you might make more money if you raise the ticket price to $20. What is the new equation for the profit? How does the graph of this equation compare to the first? Do you think that you should charge $20 per ticket? Why or why not?

d. From previous experience with concerts, it might be known that the maximum number of tickets likely to be sold, t, will depend on the ticket price, x, according to the equation $t = 1\,000 - 50x$. Study this equation. Is it a linear equation? How will its graph appear? Graph it to check your prediction.

mistake?

e. Use the two equations, $p = xt - 2\,000$ and $t = 1\,000 - 50x$, where x is equal to the ticket price, p is equal to the profit, t is equal to the number of tickets sold, and 2 000 is equal to the expenses, to investigate the likely profit for various ticket prices. What is the maximum likely profit? What ticket price would you recommend the club charge? Why?

3. For an object moving at a constant speed, the distance, d, depends on the rate of speed, r, and the time traveled, t, according to the equation $d = rt$. Investigate the fastest recorded speeds of a car, a train, an airplane, and a rocket.

a. For each record speed, write an equation that relates distance, speed, and time, then make a chart and a graph of the ordered pairs, (time, distance). What unit of time should you use? Why?

b. Are these linear functions? Why or why not? Compare the graphs. How are they alike? How are they different? Choose any distance such as 1 mile. What is the time required

for each vehicle to travel that distance? How far could each vehicle travel in some interval of time, such as 10 hours?

c. How do these various speeds affect the graphs? What kind of graph would you have for the fastest recorded speeds of a human on a bicycle, a downhill snow skier, or a human runner? How do these graphs compare with the vehicles above?

▶ GROUP PROJECT B

Investigate how to use linear graphs to find equations.

4. Temperature can be measured using either Fahrenheit or Celsius thermometers. In your group, investigate how the scales on these thermometers are related.

a. Find and study each type of thermometer. Study the scales. How are they alike? How are they different? What is the range of each? Can you tell from the scales what the freezing and boiling points of water are? Discuss your observations in your group.

b. Place both types of thermometers in a container of thoroughly-chilled ice water. Record ordered pairs of temperatures as you slowly heat the water to boiling.

c. Graph the ordered pairs (Fahrenheit, Celsius) using your data. Study the graph. What is its shape? Can you predict an ordered pair of temperatures that you have not measured? Heat or cool the water to check your prediction. After a discussion in your group, write about your results.

d. As a group, write an equation to compute values of *C* using the values of *F*. Test your equation with your data.

e. Sometimes when the temperature is given in degrees Celsius, you may want to know it in degrees

Fahrenheit. Use your data to graph the ordered pairs (Celsius, Fahrenheit). As a group, write an equation to compute values of *F* using the values of *C*.

f. Compare the two equations. How are they alike? How are they different? Compare the two graphs. Which is steeper? Why? Which intersects the vertical axis at a smaller value? Why?

▶ **GROUP PROJECT C**

Find linear equations and analyze graphs.

Assign Exercise 5 or 6 and Exercise 7 or 8 to each pair of students in your group. After completing your solutions, present your results to the other pair.

5. Which job would you choose?

- ♦ Job A: Starting salary is $20 000 a year with an annual raise of $1 200
- ♦ Job B: Starting salary is $17 600 a year with an annual raise of $1 500

a. In your group, investigate these choices. Using a calculator or computer, complete a chart for each job to show the salary for at least 10 years. At the end of 10 years, which job has the greater annual salary?

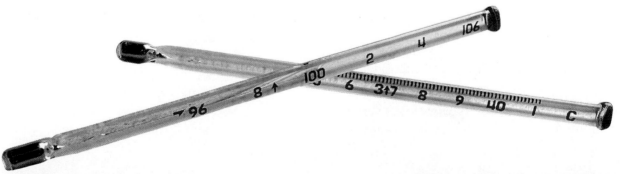

b. In what year will the salary for job A exceed $30 000? When you are earning $44 000 at job A, what would you be earning at job B?

c. Write an equation for finding the yearly salary for each job. Compare the equations. How are they alike? How are they different? How do they relate to the graphs?

d. Graph the ordered pairs (years, salary) for both jobs on the same axes. What is the shape of each graph? How are the graphs alike? How are they different? Do the graphs intersect? At what point do they intersect? What does this mean? Discuss your results.

e. Think about the accumulated earnings for each job. What are the totals after 10 years? Graph the ordered pairs (years, sum) for each job and draw the lines. If you connect the consecutive points with line segments, what geometric curve do you see? Do these lines intersect? If so, where? What does this mean?

f. In your group, make up a new job C that has a lesser starting salary than either jobs A or B, but a yearly raise greater than jobs A and B. Write an equation for job C. Compare the three jobs. Which is the "best" job after 10 years? Why? What can you conclude? Share your findings with the class.

6. The growth rates of living things vary in interesting ways. Human hair grows at the rate of about 1 centimeter every 20 days. If left uncut, it will usually grow to a maximum length of less than a meter. Study this rate using a chart and a graph. What is the shape of the graph? Write an equation based on this rate of growth. Is it a function? Why or why not?

7. The available drinking water is becoming scarce in many parts of the United States. It is important that we conserve our water. Are you wasting water in your home?

a. Sometimes water is wasted through faucet leaks. Suppose it was found that a dripping faucet lost 100 cubic centimeters of water every 5 hours. Make a chart to show the total amount of water wasted at the end of each hour for a day. Make a graph to show your data. What is the shape of the graph? As a group, write an equation to show this rate. What are the dependent and independent variables?

b. Under ordinary water pressure, if a water pipe gets a circular hole that has a diameter of 1 millimeter, up to 500 cubic centimeters of water per minute can be lost from the water system. Make another graph to show this rate. Write an equation to show this rate. Compare the two graphs. What can you conclude?

c. Extend your graphs to determine how much water would be wasted in one week.

8. Have you looked closely at the price tags on the shelves at your grocery store? Many stores provide **unit pricing**.

a. What is unit pricing? How is the unit price determined? How can you use the unit prices?

b. In your group, organize an investigation of the unit prices of different sets of products. Have each student select a product, such as frozen pizza, bread, or laundry detergent, where at least four different brands would be available. Collect pricing information for items that are the same size and type. Record the total price and unit price. Graph these ordered pairs.

c. Record the number of units and the total cost for each product. Graph

these ordered pairs on the same axes. Mark the points that show the actual size and total cost for each item. Compare the two graphs. Which items have the same value, or cost per unit? Which item is the "best buy?" Write an equation that could be used to compute the cost for some given number of units such as ounce, pound, gallon, or square foot.

d. In your group, compare the graphs. What is the shape of each? How do the graphs differ? Write a brief discussion of your conclusions.

▶ HOMEWORK PROJECT

9. Look in your school or home to find a place where water is being wasted. Study the rate of loss. Collect data that measures the rate of loss. Graph the data, and write an equation. What can you conclude? Be ready to share your results with your class.

10. Collect some data to compare prices of different sizes for the same brand. For example, pick a brand of detergent and record the size, price, and unit price for as many different-sized boxes or bottles as you can find. Do this for three or four different brands. Try to find comparable sizes. Graph and label the ordered pairs, (size, cost), for each brand. Write equations for each brand. What conclusions can you make?

SOLVING EQUATIONS

.**Y**ou have seen that many situations in the real world can be represented by linear equations. Many problems can be solved by finding particular points on a graph. Thus, if we know one of the variables, we can look at our graph and find the value of the other variable. Similarly, if we have a linear equation and pick a value for one variable, then we can solve that equation algebraically to find the other variable. In this activity, you will explore both ways.

▶ GROUP PROJECT

Investigate methods of solving linear equations.

1. Have you noticed that in a thunderstorm you see the lightning before you hear the thunder? How can you use this delay to estimate the distance to the lightning?

a. The speed of sound in air is about 343 meters per second. Use it to find out how far the sound of thunder can travel in 1 second, 2 seconds, 3 seconds, 4 seconds, and 5 seconds.

b. Make a chart of ordered pairs, (time, distance). Graph the pairs. What is the shape of the graph? Is it a function? Why or why not?

c. Connect the points on your graph. Use the graph to find d when t is equal to 20 seconds and to find t when d is equal to 5 000 meters. Here you are solving the equations graphically.

d. Write an equation involving the variable t that you could use to find d, the distance you are from the lightning. Use the equation but do not use your graph. If t is equal to 60 seconds, explain how to find d. If

d is equal to 52.7 meters, explain how to find *t*. Here we are solving the equations algebraically.

e. What is the difference between solving an equation algebraically and solving it graphically? Explain.

2. How is the speed of sound affected by a hard rain? The speed of sound in water is 1 447 meters per second.

a. Make a new chart of ordered pairs (time, distance). Use the same axes as in Exercise 1 to make a graph. Is it a function? Why or why not? Compare the two graphs.

b. Write a new equation. Compare this equation to the equation you wrote in Exercise 1. What is the coefficient of *x* and the constant in each equation? Look at the graphs of the equation. Put a value of one second into each equation. Compare the two values of *d*.

c. Pick values for *t* and explain how to

solve the equation algebraically and graphically.

d. Pick values for *d* and explain how to solve the equation algebraically and graphically.

3. Many creatures in nature are affected by seasonal changes. In the winter, bears hibernate and animals grow thicker coats. Crickets chirp more times when it is warmer. Scientists have formulated the following approximation equation, $t = c + 38$ where the temperature (degrees Fahrenheit) can be found from *c*, the number of chirps from a particular type of cricket, made in 15 seconds.

a. Use what you know about proportional reasoning to find the number of chirps in 15 seconds if 24 chirps are heard in 10 seconds, 36 chirps are heard in 30 seconds, 18 chirps are heard in 5 seconds, and 72 chirps are heard in 24 seconds.

b. Now find the temperature using the values you found in Exercise 3a and the formula $t = c + 38$. Make a table relating the number of chirps to the temperature.

c. Graph the ordered pairs, (c, t). What is the shape of the graph? Is it a function? Why or why not? Are there any values that will not work in this equation? Why or why not?

d. Select values for c and explain how to algebraically solve for t. Select values for t and explain how to solve for c.

e. Check your algebraic solutions on the graph you made above.

f. Research to find how some other type of cricket, animal, bird, or insect might be affected by temperature. Find data or an equation that can be used to describe this effect. Make a chart and a graph. Write two problems where an equation must be solved. Exchange problems with other groups.

4. In Activity 5, Exercise 4, you found two equations for converting between temperatures in Fahrenheit and Celsius. One equation is $F = 1.8C + 32$.

a. Explain how to find C if $F = 98.6°$, the normal human body temperature.

b. The normal body temperature of a dog is about 101.5°F, a snake is about 95°F, a whale is about 98.6°F, a duck-billed platypus is about 86°F, and a robin is about 41°F. Explain

how to find each body temperature in degrees Celsius.

▶ HOMEWORK PROJECT

5. In this problem, you will investigate the amount of money earned by a carpenter and an assistant. The carpenter earns $35 an hour and her assistant earns $25 an hour. Every day the assistant starts work at 7 A.M. and the carpenter starts at 9 A.M.

a. Make a table showing the time of day, the hours worked so far by each person, and the money earned by each person.

b. Graph this data for both the carpenter and the assistant.

c. Write an equation for each worker.

d. Find when the carpenter and the assistant have earned the same amount of money, graphically and algebraically. Show and explain your work.

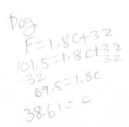

Dog
$F = 1.8C + 32$
$101.5 = 1.8C + 32$
$ -32 -32$
$69.5 = 1.8C$
$38.61 = c$

EXPLORING NONLINEAR FUNCTIONS

Many situations involve functions that are not linear. You have already seen many examples of **nonlinear functions**.

In this activity, you will investigate some of these types of functions. You will study their graphs and equations and see how to solve such equations.

▶ PARTNER PROJECT A

Investigate nonlinear functions by extending patterns.

1. Look at the figures below.

PORTFOLIO BUILDER

i. $y = x^2$

ii. $y = x + x$

iii. $y = x^2 - 1$

(no line segments)

iv. 2, 8, 18, 32

v. 5, 8, 13, 20

   a. In each of the patterns above, sketch the next figure on pages 104, 105, and 106 of your Portfolio Builder.
   b. List these terms and the next four terms of each on pages 104, 105, and

106 of your Portfolio Builder.
   c. In each, describe the patterns that you notice.
   d. Continue each chart to include the *n*th term.

2. Graph each group of patterns listed below on separate axes. Label each graph.
   a. Exercise 1, patterns i and iv
   b. Exercise 1, patterns i, iii, and v
   c. What shape are these graphs? Explain the similarities and differences you see in the graphs.

3. In Activity 2, you investigated families of linear equations. Here are several new families of equations where *x* is any real number.

Family 1

$y = 0.5x^2$

$y = x^2$

$y = 2x^2$

$y = 4x^2$

Family 2

$y = -0.5x^2$

$y = -x^2$

$y = -2x^2$

$y = -4x^2$

Family 3

$y = x^2 + 1$

$y = x^2 + 4$

$y = x^2 - 1$

$y = x^2 - 4$

Family 4

$y = (x + 1)^2$

$y = (x - 1)^2$

$y = (x + 3)^2$

$y = (x - 3)^2$

Family 5

$y = (x + 1)^2 + 4$

$y = (x - 1)^2 - 4$

$y = (x + 3)^2 + 1$

$y = (x - 3)^2 - 1$

PORTFOLIO BUILDER As you go along in this exercise, keep a record on pages 107 through 111 of your Portfolio Builder of sketches, charts, ideas, and answers to questions. You will be using these items to continue the report for your portfolio.

a. In your group, study and discuss each family. How are the equations within each family alike? How are they different?

b. On a computer or graphing calculator, compare the graphs within each family. How are they alike? How do they differ?

c. For each equation within the family, make a table to find the coordinates of five ordered pairs. Remember to use both positive and negative values for x. To understand the relationship, study how the value of y varies as the value of x changes.

d. Graph each family of equations on the same set of axes so that you can compare them more easily. Label each graph with its equation. Discuss your graphs with the class.

e. Now compare the families for differences and similarities. How are they alike? How do they differ? How does a change in the equation between families seem to affect the graphs? Record your answers on page 112 of your Portfolio Builder.

f. On page 112 of your Portfolio Builder, explain how family 5 relates to families 3 and 4.

g. Write a paragraph of your conclusions about these families on page 112 of your Portfolio Builder.

4. COMPUTER ACTIVITY Complete the BASIC program below to make a table of values for the first family above.

10 REM Comparing Algebraic Expressions

20 PRINT " 0.5x² x² ___ ___"

30 FOR x = 1 TO 20 STEP 0.5

40 PRINT 0.5 x^2; ___; ___; ___

50 NEXT x

60 END

RUN the program and use the results. What does the program choose x to be? Change the program to find the tables for each of the other families.

▶ HOMEWORK PROJECT

5. Write a new family of nonlinear equations that is similar to families 1 through 5 in Exercise 3. Investigate how the graphs of your new family behave. Write a brief report of your results.

▶ PARTNER PROJECT B

Investigate different types of curves.

6. Refer to Unit 6, Activity 6, where you used waxed paper folding to make envelopes of several different curves.

a. Which of the envelopes appear to show a curve whose shape is like the graphs you produced in Exercise 3? Why? What is the name of this curve?

b. Have each member of your group fold paper to form the envelope of a parabola or use one from your portfolio. Mark and label 8-10 points of the curve. On a sheet of graph paper, mark the *x*- and *y*-axes. Place the wax paper parabolas on the graphing axes. You will want to discuss how to place the curve on the coordinate system. Record the ordered pairs (x,y) for each of the points marked on your folded curve. Save this waxed paper folding for your portfolio.

c. Study the ordered pairs. Compare the values to the charts you made for the families above. Do any of the expressions "fit" your folded curve? If not, can you modify an expression to "fit?"

d. By now you may know that the expression will change if you slide the folded curve up or down, or if you turn or flip it. Place the folded curve in a new position on the coordinate system. Record the ordered pairs (x,y) for the marked points. Try to find an expression that will "fit" this graph.

7. As you climb from sea level to a high mountain top, the atmospheric pressure reduces, the air becomes thinner, you have more trouble breathing, until at some point you require an oxygen tank in order to continue breathing. As you dive deeper under water, pressure increases, and you might feel a "squeeze" if air trapped in your body cannot equalize.

a. What do you think happens to the bubbles of air that escape from a diver's mouthpiece? Are they the same size as they break the surface of the water? Why or why not? Explain what you think happens.

b. The following is a graph of the volume of air as it relates to atmospheric pressure in a freshwater lake.

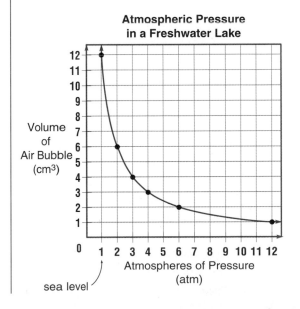

What is the shape of the graph? Is it a function? Why or why not? What is the independent variable? What is the dependent variable?

c. What is the volume at sea level? What is the volume when atmospheric pressure (atm) is 6?

d. Use centimeter cubes to model these two volumes. Compare these volumes.

e. Find the atmospheric pressure when the volume is 4 cm^3.

f. If one atm of pressure supports a column of water 10.4 meters high, how deep is this diver when the air bubble is 2 cm^3?

8. Refer to Unit 6, Activity 6, where you used waxed paper folding to make envelopes of curves. Which of the curves resemble the curve in Exercise 7? If you saved the curve in your portfolio, take it out and examine it. Otherwise, remake the folding of that curve on waxed paper.

9. On your calculator you have a $\sqrt{x}$ key. How does this **square root function** behave?

a. Investigate $y = \sqrt{x}$. Make a chart of ordered pairs $(x, \sqrt{x})$. Graph the ordered pairs. What is the shape of the graph? For what values of x is the $\sqrt{x}$ defined? Why?

b. Create a family of equations for the square root function. For example, you might study

$y = -\sqrt{2x}$
$y = \sqrt{3x}$
$y = 2\sqrt{x}$
$y = 3\sqrt{x}$
$y = \sqrt{(x - 1)}$
$y = \sqrt{(x + 1)}$.

Graph the family on the same axes. Compare the graphs. What can you conclude?

c. Is the graph of $y = \sqrt{x}$ similar to any curve you have studied in this activity? Which curve and why?

▶ **GROUP PROJECT**

Investigate applications of curves.

In your class, divide up the following exercises. Each group should choose one of these and prepare an oral report for the class. When preparing your report, include graphs, tables, explanations, notes for the chalkboard, demonstrations, and handouts, where appropriate.

10. In Activity 3, Exercise 2, you explored how the equations for ages of your family members might be written and varied.

a. How does the product of your age and your mother's age vary? Write an equation for the product, given any number a for your age. Make a chart and a graph to study this. When is the product an even number? An odd number? A prime number? Equal to 100? Equal to 1 000?

b. On the same axes, write equations and make graphs for the products of your age and each family member. Compare the expressions and their graphs. What can you conclude?

11. How does the handshake problem relate to the number of diagonals in a polygon?

   a. Review your work from Unit 6, Activity 1, Exercise 4. Study the data and the graph for the number of diagonals in a polygon of *n* sides. Does it appear to be a function? Why or why not?

   b. Review your work from Unit 1, Activity 5, Exercise 14. Study the data and the graph for the number of handshakes with *n* people. Does it appear to be a function? Why or why not?

   c. Compare the two graphs. How are they alike? How are they different? Write an equation for each graph. Compare the equations.

12. Investigate how a playground seesaw behaves.

   a. Suspend a meter stick with a string at the 50-cm mark. Make sure the meter stick balances. If not, tape a paperclip to the stick so that it does. Then use large paperclips to hang weights, such as washers, on each side to simulate persons of different weights sitting at various places on a seesaw. Keep one side fixed and vary the weight and distance on the other side. Experiment with different numbers of weights on each side. How do you get it to balance?

   b. Organize an investigation to study how weights and distances from the center can vary when the seesaw is balanced. Set up several different cases, including "children" of different weights. Investigate what happens when two "children" of different weights and distances on one side balance one "child" on the other side. Record and study the data.

   c. Graph the ordered pairs (weight, distance). What is the shape of the graph?

13. Investigate how a playground swing behaves.

   a. Suspend a weight on a string. Use a stopwatch to measure the time, called the **period of the pendulum**, for 1 swing over and back. To get a

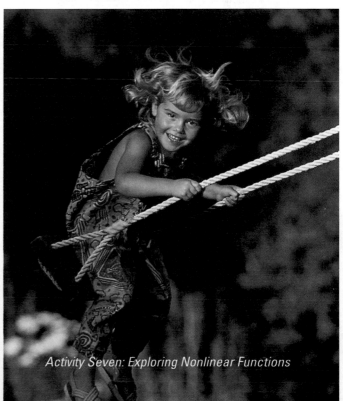

better measure, you might time 10 swings, then divide by 10. What affects the period? Is it a different weight? A different starting height? Think about when you have been on a swing.

b. Organize an investigation of the period of a pendulum. Take measures to obtain at least 8 ordered pairs, (length of the string, period). From the data, how does the period seem to be affected by the length? Explain.

c. Graph the data. What is the shape of the curve? How does it compare to other graphs of curves studied in this activity? What kind of equation would you write for the period of a pendulum?

14. In many situations, you have to compare the affects of different speeds.

a. Suppose that Enrique leaves from his house on his bike, riding 20 km/h. One hour later, his friend, Alphonso, leaves Enrique's house on his bike to catch up with Enrique, riding along his route at 30 km/h. Write equations for each rider to show the distance, d, found from the amount of elapsed time, t, measured from when Enrique starts.

b. Use the equations to find the elapsed time when Alphonso catches Enrique. Check your result. Clearly explain your solution.

c. Assume that when Alphonso catches Enrique, he slows down to match Enrique's pace (20 km/ h). Their friend, Harris, also leaves Enrique's house 30 minutes behind Alphonso, riding along the same route at 35 km/h. Write an equation to show Harris' distance using the time, t, measured by Enrique's ride.

d. Use the equations to find when Harris will catch Enrique and Alphonso. Check your result. Explain your solution.

CHECKING YOUR PROGRESS

In this activity, you will demonstrate some of the ideas you have learned in this unit. In Part I, you will explain what you have learned about patterns in this unit. In Part II, you will work with another student to develop a joint report related to some problem situation. In Part III, you will complete your work on the Unit Problem.

PART 1 INDIVIDUAL ASSESSMENT

1. Write a letter to an adult in your home that explains what you have been learning about patterns and how they relate to mathematics. In that letter, include the following.
   a. A drawing and complete explanation of a pattern that was in this unit
   b. A pattern of your own, using drawings or blocks, that fits a related linear or nonlinear equation

   c. For the pattern you develop in Exercise 1b, make a table, find the formula for the nth term, graph the pattern, explain the shape of the graph, and tell whether your graph is a function. Explain why or why not.

2. Use pages 97 through 101 in your Portfolio Builder and the graph below.

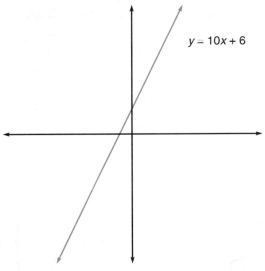

$y = 10x + 6$

Write an equation of a line in the same family that would be:
a. steeper looking
b. flatter looking
c. horizontal
d. vertical
e. same steepness but higher on the axes
f. same steepness but to the right on the axes.

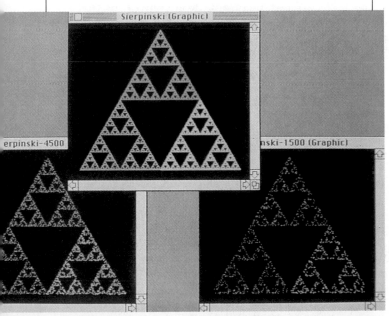

3. Use pages 107 through 111 in your Portfolio Builder and the graph below.

PORTFOLIO BUILDER

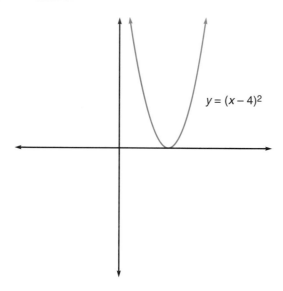

$y = (x - 4)^2$

Explain how the following equations will affect the graph.

a. $y = (x - 4)^2 + 1$
b. $y = 2(x - 4)^2$
c. $y = 0.5(x - 4)^2$
d. $y = -1(x - 4)^2$

PART II   PARTNER ASSESSMENT

With your partner, select one of the following items. Develop a written report of your results to be submitted to your teacher.

4. Have you ever traveled to a foreign country? Other countries have their own money system. As a traveler, you will want to know how to convert between U.S. dollars and the currency of the country you are visiting.

a. In the business or financial section of the newspaper, find the current exchange rate for the British pound sterling. What does this rate mean? Use this rate to make a chart and a graph for ordered pairs, (dollars, pounds). What type of function is this?

b. Write an equation to find pounds from dollars. Check the equation for values in your chart.

c. Use the equation from above. Solve the equation to find dollars from pounds. Show the steps of your solution. Check your new equation.

d. Find the exchange rates for Canada, Germany, Japan, Mexico, and at least one other country you would like to visit. What is the name of the monetary unit in each country? Discuss what these rates mean. For each country, write equations to convert both ways between U.S. dollars and the local money.

5. On Interstate 5, a California Highway Patrol Officer sits under an overpass with his radar gun pointed at the oncoming traffic. A Lamborghini is speeding, and the officer clocks the driver traveling 80 miles per hour. By the time the officer puts his equipment away and accelerates to a speed of 100 miles per hour, the Lamborghini is now 1.6 miles ahead of the officer. When will the police car overtake the speeder? Fully explain your work using tables, equations, and labeled graphs.

6. How long does it take to stop a car? For an alert driver in a car with good brakes driving on a dry, paved road, you need to consider both the distance the car will travel while your body reacts and applies the brakes, and the distance the car will travel after the brakes are applied. Two formulas are used here. The formulas can be combined to find the total distance traveled by the car.

$R = 1.1s$ feet, where s is the speed of the car and R is the reaction distance.

$B = 0.05s^2$ feet, where s is the speed of the car and B is the braking distance. The total stopping distance is $D = 0.05s^2 + 1.1s$.

a. Investigate stopping distances. Develop a table and a graph to show what happens for various speeds.

b. Explain what happens as you change from 20 mph to 40 mph. Compare this with going from 20 mph to 60 mph. In each case, graph the ordered pairs for (speed, reaction distance) and (speed, braking distance). Discuss these graphs.

c. Make a number line display of a road to show with a scale drawing where the car would stop for various speeds. For each stopping distance, tell some familiar length such as a city block or a football field.

d. The reaction time for an average driver is about 0.75 seconds. Use the equation $d = rt$ to show why the reaction distance in feet is 1.1s. [Hint: When the speed is miles per hour and the time is in seconds, how can you find the distance in feet? Use ratios to change the units.]

PART III   UNIT PROBLEM

Complete your cooperative and individual work on your unit problem. Write and present your report.

ENDING UNIT 7

 Now that you have completed Unit 7, refer to page 113 of your Portfolio Builder. You will find directions for building your personal portfolio from items you completed in this unit.

PORTFOLIO BUILDER

GROWTH AND

DECAY

TRANSCEIVERS,AIRBORNE, CONTR W/INDIAN GOVERNMENT
DEVELOPMENT OF AN/TRM SWEEPGEN, USN
DEVELOPED PREPROGRAMMABLE RECORD CHANGER SYSTEM
DEVELOPED SIMON SINGLE CHIP-UP GAME
DEVELOPED COMPUTERPERFECTION GAME
DEVELOPED AMAZATRON HAND HELD GAME
DEVELOPED VCR-GAME WITH VIDEO AND AUDIO BRANCHING
CONCEPT FOR VR-CONTROLLED ANIMATED FIGURE VCR-GAME
BUILT 1 PLANE, P/O PROPOSAL TO USN, JOHNSVILLE PA COMPUTER
CONCEPT. P OPOSAL TO USN AIRBO
BUILT 1ST   PROPOSAL
DEVELO
BUILT
BUILT   CE UNIT
BUILT   E GAMES
BUILT

            TENTS (R.H. BAER   E
                                          SYSTE
                114   GENERATOR
            119844   —— 22 AUG 80 PROGRAMMABLE RECORD CHANGER
            4153821   —— 10 OCT 78 PHONOGRAPH PHONOGRAPH DEVICE
            4166621   08 MAY 78 OPTICAL SCANNER MAKING MEANS
            4200708   —— 04 SEP 79 PROGRAMMABLE RECORD CHANGER
            4216965   —— 10 JUN 80 MICROCOMPUTER CONTROLLED GAME
            4240638   —— 12 AUG 80 MICROCOMPUTER CONTROLLED GAME
            4768967   —— 23 DEC 80 MICROPROCESSOR CONTROLLED ELEC GA
            4846693   —— 22 NOV 88 INTERACT VIDEO APPARATUS W/AUDIO
            4487526   —— 11 JUL 89 VIDEO BASED INSTRUCT & ENERT SYST
            3320604   1921 17 OCT 89 ENCODING OF AUDIO AND DIGITAL SIG.
            3629939   1972A 10 FEB 65 INDICATOR ANNOUNCING SYSTEM
            3433442   2145 18 MAR 69 MULTILAYER CORE MEMORY PROCESS
            3611321   2303 04 APR 69 PARACHUTE DEREEFING SYSTEM
            3599221   2560 10 AUG 71 MEMORY DEVICE & METHODS
            3658285   2837 25 APR 72 RECORDING CRT LIGHT GUN METHOD
            3728480   2401D 17 APR 73 TELEVISION GAMING APPARATUS
            3737566   2401C 05 JUN 73 TELEVISION GAMING APPARATUS
            3829095   2401B 13 AUG 74 TELEVISION GAMING AND TRA
            RE28598   2837   TE

$m^2 - 2m = 35$

$3x^2 + 14x =$

ach quadratic equation.

7. $2y^2 + 3$
10. $3k^2 =$

$- t - 15 = 0$
$4a^2 + 8a = 0$

13. m
16. $14x = 5$

uadratic equation.

$= -8k$

What can you do with a television set other than watch it? Well, in 1966, Ralph H. Baer had an idea. Play games! He began spending all of his free time working on his idea. On April 25, 1972, patent number 3 659 285 was issued "... for the generations, display and manipulation of symbols upon the screen of television receivers for the purpose of playing games..." Video games had arrived!

Born in Germany in 1922, Baer and his family immigrated to the United States in 1938 to escape the Nazi regime. As a naturalized U.S. citizen, he enlisted in the army during World War II. While stationed in England, he finished a correspondence course in algebra, which helped prepare him for post-war careers as an engineer and an inventor. Mr. Baer cites his proficiency in mathematics as a key to his success.

Mr. Baer has nearly 100 patents relating to video games. The first one was licensed by Magnavox in 1972 for Odyssey 100 - the first home video game. He has also had success in the portable electronic game market, developing Milton-Bradley's Simon among others.

Today, in his 70s, he is still a full-time independent electronic-toy-and-game inventor.

ABOUT THIS UNIT

Have you ever watched something grow and change? Have you ever returned to a place and seen hills or beaches that have gotten smaller? Perhaps you have seen how some animal, such as a pet, has grown from a baby to an adult. Plants often grow rather quickly to reach maturity in a single growing season and then wilt and die. Or, maybe you know how investments or savings can grow from earning dividends or interest. Have you ever heard of investments called growth funds? Sometimes, however, the change is a decrease in the size or amount, and this is often referred to as decay. Have you ever heard of some investments that decrease?

There are a series of mathematical and scientific investigations within this unit that you will need to set up and monitor. You have completed many mathematical investigations throughout this course. Therefore you will be expected to be able to set up the investigations that you conduct in this unit, as well as follow through with data gathering and write-ups.

On this first day of the unit, read the projects on pages 254, 256, and 273 and begin setting up your experiments and collecting the necessary data. As you read each project, be sure to make a list of all materials you think you will need, write a plan of action that includes a checklist for times, and design a lab sheet for each investigation to use to record your data.

The experiment on plant growth should have been set up 2 or 3 weeks ago. If it was not started at that time, be sure to start it today and begin gathering your data as soon as growth is observed.

In this unit, you will use mathematics to describe different kinds of growth and decay. You will investigate different situations to observe patterns of change. You will also use the mathematical ideas you have learned in this course, such as graphs, functions, variables and equations, large and small numbers, data organization and analysis, probability, geometry and measurement.

▶ UNIT PROBLEMS

1. HOUSING

The growth of a city can lead to a demand for housing. The rate at which new housing units are constructed will go along more or less with the rate of growth in the population.

a. Find out about the population growth and/or decay, of your city or some city of your choice. Obtain population data for each census as far back as possible. Graph the data. Determine what factors may affect the rate of growth or decay of your city. Explain. What would you predict for the next census? Why?

b. Find out about the construction of new housing units in your city. Determine what types of units and how many of each type are built each year and graph the data. Find factors that affect the construction of new housing units. Predict housing construction for new units for the next few years and explain your prediction.

c. Insulation can hold in, or keep out, heat at a rate determined by its thickness. The two kinds of insulation most often used are cellulose-treated fiber and fiberglass. Make a list of thicknesses and insulation effectiveness values of each thickness. The unit of measure used to determine insulation effectiveness is called R-value. Explain what R-value means.

d. Study how sound can be affected by the thickness of wall materials. Use graphs and exponential expressions to indicate the decay of the level of sound as the thickness increases. Find the thickness of walls for an average level of a sound. If you live next door to someone with unusually loud music, how thick should the walls be? Explain.

e. Compare the pattern of growth for population and for housing in your city. Explain any patterns you notice. Compare these to other cities, regions, or nations. Make a prediction about the availability of housing in your city over the next few years and explain how you made your prediction.

2. WEALTH

The level of wealth in your city can increase or decrease due to many factors.

a. Study the wealth of people in your city or some city of your choice. Try to obtain data from the past that would show the level of wealth compared to other cities, regions, or a national average. Graph the data to show how the level has changed. See if the finances and spending of the city government is related to the level of wealth of the citizens. Explain the relationship.

b. Compare the size of the population to the level of wealth. Explain how they are related.

c. Explain the wealth trend for your city. Use it to predict the future wealth. Explain how wealth patterns should or could affect the finances of city government.

3. HEALTH

The growth of a city can lead to increased demands for health services. Some of these needs must be met through public support that can require more public revenues.

a. Study the health services in your city or some city of your choice. Find information that will let you show how these demands have grown, perhaps by decades, during the past century. Report on how many hospitals, public health centers, private clinics, doctors, nurses, pharmacies, and ambulance services there are and how these numbers have changed.

b. Compare the growth of health services with the growth in population. Try to relate this data, so you can conclude how many additional citizens seem to require a new hospital or another doctor. Explain how these new health services are financed.

c. Predict future needs for health services in your city. Explain how you made your prediction.

4. POPULATION

The population of your city, or some city of your choice, has probably changed over the past century. Study these changes in more detail.

a. Find information about the population changes in your city. Specifically, try to obtain data to show such breakdowns as age, gender, ethnic background, economic level, educational level, number of children, or profession.

Explain any changes in the nature of the population shift. Investigate how the changes in the size of the city have affected the demands on communal services, such as schools, public transportation, public low-cost housing, electric power, natural gas, water, sewer, telephone, or cable TV. Try to find out how many new citizens lead to another increment of service. Explain how these communal services have been paid for. Determine and explain any effect on taxes.

b. Predict and explain any need for specific new communal services that will be needed in the next decade or two. Explain how the services can be supported.

CHANGES

We live in a world that seems to be in constant change. Mathematics can help us to model and understand changes. In this activity, you will explore situations that involve change. You will learn how mathematicians analyze these situations by collecting, organizing, and looking at data to see any trends.

▶ GROUP PROJECT A

Investigate situations of growth and decay.

1. **PORTFOLIO BUILDER** What things in life grow and/or decay? In your class, make a list of those things. Next to each item, list the possible factors that might affect each one. Put these lists on page 114 of your Portfolio Builder, so that you can add to them throughout this unit.

▶ GROUP PROJECT B

Investigate growth and decay of plants.

2. In this activity, you will study how a plant changes as it grows and decays. This activity will take approximately one month to complete. You will need to gather and record data daily.

   a. In your group, discuss how you will study the growth and decay of plants. Determine what data you should collect. What factors can affect growth and decay, and how will you control or manipulate these factors? Scientists choose an independent variable and look at the results when they manipulate that particular variable. Decide on some variables that you will observe and choose some factors affecting growth that you will control or manipulate such as the amount of moisture, the amount and type of fertilizer used, the amount of sunlight, the type of soil, the time of planting, and the orientation of the plant to the sun.

   b. As a group, decide on a plan for your investigation of plant growth. You may wish to add some other items to your own study. Be sure to record your plan so that you are consistent in its follow-through.

   c. Use fast-growing seeds such as green beans, black-eyed peas, pumpkin, squash, watermelon, soy, or bamboo shoots. Because plants often show great variation in their development, collect data from as many seeds of

the same kind as possible. Each member of your group should select the same type of seed, planting at least three seeds per person. Soak the seeds in water for 24 hours, or until they germinate. Germination takes the shortest time in a dark, humid place. Wrap the seeds in moist paper towels and put them in an open plastic bag.

d. Place the bag in a dark cupboard or under a cardboard box. Carefully inspect your seedlings each day. After about 5 days you should see sprouting. Which appear first, roots or stems? Why? Once your seeds have sprouted, plant them in a paper cup with potting soil. Use your finger to poke a hole in the soil about one inch deep. Be sure to place the root down and the sprouting part upward. To see the root growth without damaging the plants, you may want to plant your sprouted seeds in a glass jar or clear plastic cup.

e. Measure or count to obtain the needed data each day. Consider recording any or all of the following:

 ◆ length of the root
 ◆ length of the stem
 ◆ number and lengths of root branches number and lengths of stem branches number of leaves
 ◆ size of leaves
 ◆ amount and weight of water used
 ◆ plant weight for each day

 ◆ which day you noticed the first green coloring and what this signifies
 ◆ if you apply plant food, record the amounts and dates

Take at least two photographs of your plant, one near the beginning and one near the end of this experiment.

f. Make graphs to show your data. Identify the independent and dependent variable in each comparison. Explain how graphs can help to show the rates of growth. For example, do all seeds of one type show the same growth pattern? Is the total daily growth of the plant a constant? How does the growth of different types of seeds vary?

g. One useful way to analyze the data is to compute ratios. For example, compute the ratio of the length of the stem to the length of the root for each successive day. How does this ratio change? Does it ever equal, or exceed, 1? How big could this ratio become? Why? Consider other ratios that may help you to analyze your data on plant growth.

h. After many days or weeks, your plants may decay and die. Study the changes that occur as a plant decays. Uproot one of your plants and wash off the soil. Gather data that can be used to describe or picture the decay process such as measuring the length and weight, and counting the roots, branches, or leaves.

i. Write a report of your findings. Be sure to include the ideas you investigated, the recording sheet you used, your data, any graphs or charts, and a write-up of your conclusions.

▶ HOMEWORK PROJECT A

3. Have you ever looked closely at the leaf of a tree? Investigate how the sizes of leaves vary, and, knowing the size, how useful the information is.

   a. Select a tree or shrub, perhaps in your yard or in your neighborhood, that has leaves with a regular or smooth outline. Collect a sample of about 20 various-sized leaves from the same plant. Measure and record the width and length of each leaf to the nearest millimeter.

   b. Study the data for patterns. What can you conclude? Graph the ordered pairs, (length, width). What can you see from the graph?

c. Use the data to compute the ratio, length to width, for each leaf. Find the mean ratio for your data set. What special meaning does this ratio have for the graph you made above? When you return to class, compare your results with others in your group.

d. Recall in Unit 2 how you made a line plot to show how the values in a data set were distributed. Make a chart of the mean ratios found by each student in your class. Make a line plot from the chart.

4. Investigate how the ratio of length to width varies during different stages of the same leaf's growth. Select a few very young leaves on a plant, and carefully "tag" them so you can identify them. Gently measure the length and width without damaging the leaf. At regular time intervals, measure and record these values for each leaf in your sample until it reaches its mature size. Graph the ordered pairs, (length, width). Find the ratio of length to width for each and find the mean ratio. What can you conclude? Compare your results with others in your group.

▶ PARTNER PROJECT A

Investigate human fatigue as a decay process.

5. Any kind of work causes fatigue. When you get tired, you can rest awhile, recover, then go on. Fatigue can cause a decline in performance, a change we

can think of as decay. Explore how muscle fatigue affects your performance.

a. Study muscle fatigue with a simple exercise of opening and closing your fingers on one hand by following these rules.

- ◆ Have one person do the following exercise. Sit with your right arm extended straight, palm up, on the desk top in front of you with your fingers together and your hand closed. Open and close your hand as fast as you can, being sure that your open fingers touch the desk and your closed fingertips touch the palm of your hand each time.

- ◆ Have your partner use a stopwatch to time the exercise and record the number of completed hand openings and closings. The exerciser will continue for 30 seconds and then rest for 30 seconds. The exercise sequence will be repeated four more times by the same person. After a 60-second rest, repeat the sequence with the left hand. Be sure to record the left-hand exercise separately from the data collected from the right hand.

- ◆ Exchange roles and repeat the exercises so that each partner has their own set of data.

b. Make a chart to record that numbers for each 30-second period (not counting the rest periods) and for each hand.

c. Graph the data for each hand on the same axes. Identify the independent and dependent variables and draw the curve of best fit. Explain the trend of performance as an example of growth or decay. If the exercise caused fatigue, did you recover quickly? Which hand showed the better performance? What might explain any differences? Locate the muscles that you used. Do you think they are the same for both opening and closing your hand? Explain.

d. From the pattern of change found in the graphs, it might be possible to predict, or extrapolate, beyond the data. Use the graph to estimate your performance in a sixth 30-second period.

e. How might the results differ if the 30-second rest period was removed?

▶ **HOMEWORK PROJECT B**

6. Repeat the experiment in Exercise 5 at home with a family member to help you. Use six 30-second periods without stopping. Graph the results and compare the pattern of change to the first experiment and your prediction in Exercise 5e.

7. Investigate muscle fatigue resulting from an exercise of your choice. Collect, graph, and interpret the data.

▶ PARTNER PROJECT B

Investigate the growth and decay of a person's weight.

8. One of the familiar ways we might experience gains or losses is with our own body weight. The average male needs about 2 800 calories daily, while the average female needs about 2 200 calories daily. To understand how you gain or lose extra pounds, a person with an average metabolism will gain a pound of weight from consuming 3 500 calories more than normal, or lose a pound by eating 3 500 calories less than normal.

   a. How many calories do you typically consume in one day? Write down everything you eat during a 24-hour period. For each item, make note of the quantities you consume. For example, do not write down one bag of chips. Instead, record 1 six-ounce bag of potato chips. If there is a calorie count on the bag, be sure to write it down. You may want to keep track of what you consume over several days and compute an average.

   b. Do you, or someone you know, want to lose or gain weight? How much, and how quickly? Consider the fictional people listed below. Choose one of these people and create a meal plan to help them achieve their goal.

Assuming the person is moderately active, you can find out how many calories he or she needs to consume daily to maintain that weight by multiplying the person's current weight by 15. List their daily menus and graph their progress. Be sure to indicate the independent and dependent variable.

Andy: Your current weight is 150 pounds. You need to gain 10 pounds to be able to wrestle in the next weight class two weeks from today.

Beth: Your current weight is 162 pounds. You would like to lose 32 pounds by next summer.

Carl: You used to weigh 190 pounds, just lost 20 pounds, and you have 10 more to lose by next month.

Diane: You just found out that you have diabetes and that you will be a lot better off if you can lose 20 pounds and get down to your ideal weight of 130 pounds.

Eddie: You weigh 275 pounds and the doctor says you need to lose 100 pounds to reduce your chance of a heart attack.

Fran: You are 30 pounds under your ideal weight of 120 pounds because you just got off a terrible starvation diet.

Gabby: You weigh only 98 pounds. Your coach has told you that if you can gain 12 pounds, you will be a lot more powerful on the court.

Henry: You weigh 325 pounds and your team is fining you $1 000 a day for being overweight. You can weigh no more than 300 pounds and play professionally. A starvation diet will leave you in no condition to play.

Inez: You would like to lose 10 pounds for your wedding. Your current weight is 143 pounds.

Jack: Your wife has agreed that she will buy you a whole new wardrobe if you can lose 2 sizes. This means trimming about 30 pounds off your current weight of 190 pounds.

Karen: You are sick and tired of being called a bean pole. Your current weight is 82 pounds and your ideal weight is 110 pounds.

Lola: You just got a role in a new movie provided you gain 50 pounds for the character. Production starts in 10 weeks. You currently weigh 134 pounds.

Mike: Last week, security picked you up because you didn't have a hall pass. You figure if you gain 20 pounds, you may look more like the teacher you are than a student. Your current weight is 114 pounds.

Nancy: You just got back from being lost in a jungle where your plane went down while on assignment. You've lost 33 pounds. All of your clothes hang on you and you can't afford to buy a whole new wardrobe. You figure that if you can gain only $\frac{2}{3}$ of your weight back you'll be at your ideal weight of 120 pounds and your clothes will look nice.

▶HOMEWORK PROJECT C

9. Make up a healthy meal plan for yourself. If you want to gain or lose weight, incorporate that into your plan by increasing or decreasing your calorie intake. Make a graph of the progress you would make if you followed your ideal plan. Keep in mind that rapid gains or losses are usually not permanent changes in weight. Without strict supervision by a physician, you should eat at least 1 000 to 1 200 calories each day.

10. You can also change your body metabolism by increasing your daily activity through exercise or strenuous work. The chart below shows the calorie-burning potential of several activities.

DAILY ACTIVITIES	
ACTIVITY	**CALORIES BURNED PER HOUR**
Bicycling at 8 mph	400
Chopping wood	450
Digging in a garden	516
Relaxed swimming	250
Fast swimming	640
Hiking hills	490
Mowing the yard	420
Paddling a canoe	580
Raking leaves	282
Walking at 4 mph	348
Weeding	336

Using your average daily food intake, add exercise or strenuous activity to your daily routine and compute the change that this would cause if you did not increase your calorie intake. Also compute how much extra food you would need to consume to balance the extra activity and maintain your weight. Make graphs to show the daily weights that should theoretically result from your plan over a period of time.

11. If you adopt any of your plans and follow them to reach your "ideal" or goal weight, keep records of your

results and compare them to your theoretical results. If you feel comfortable to do so, share the results with your class.

▶ PARTNER PROJECT C

Investigate situations of growth and decay.

12. A bouncing ball illustrates the idea of decay. Investigate the decay in the rebound height of different types of balls as they bounce. Drop the ball from a height of, say 2 meters, and observe the height of each successive bounce. Make a graph of the ordered pairs, (number of bounces, height) for each ball. Study the graphs to describe the decay.

13. Change makes life interesting. It is especially true when the change is unpredictable. The idea of unpredictability is basic to communication. In writing or speaking, highly predictable information can become boring. Yet, we usually don't want to give the maximum amount of information possible, so most writing has a large amount of redundancy to lead the reader smoothly through the ideas.

a. The redundancy of written English can be studied easily. Write or choose a logical sentence of at least ten words. Flip a coin to either delete or skip each successive letter. Give the sentence with spaces showing the deleted letters to a sample of people. Ask them to fill in as many letters as possible. Find a percent correct for each person. Did you get what you expected? Why or why not?

b. With the same sentence, roll a number cube with each letter. If you roll a 1, 2, 3, or 4, delete the letter. What portion of the letters would you expect to delete? Why? Again, ask several other people to fill in the deleted letters. How does their performance differ from the first group? Why might this be?

c. Repeat the experiment with a paragraph of at least five sentences. Compare the results for two groups given about half versus two-thirds of the letters deleted randomly. Describe the differences between the two groups.

CAN YOU BE A MILLIONAIRE?

Can you become a millionaire? How long would it take, and how much would you need to invest?

Some things can grow or decay slowly at first, but then very rapidly. In this activity, you will investigate how compounding causes a nonlinear change. You will see how exponents can be used to model multiplicative growth.

▶ **GROUP PROJECT A**

Investigate the effects from compounding interest.

Because of computers, most investments today earn interest or dividends on a frequent, sometimes even daily, basis. What this means is that the earned interest is credited to your account, so that this

amount also begins to earn interest. This is called **compounding**, and you will see that it can greatly affect the rate at which your account grows. To understand how your investments might grow, we will first work with simple interest computed each year.

1. Consider an interest rate of 3%, or find the current interest rate paid by a local bank. Assume that you can deposit or invest $10 000.

   a. **Simple interest**, interest paid only on the original principal, is paid by the formula $I = prt$, where I is the interest, p is the principal or the amount of money invested, r is the interest rate , and t is the time the money is invested. Use this formula to compute the interest for one year, then compute the interest for each of

20 years. Make a table to show the year, total interest, and total money for each of the 20 years. Then graph the ordered pairs (years, total money).

b. Normally, the bank does not send you the interest money, but rather deposits it into the account where it becomes part of the amount used to compute the interest earned in the next period. We refer to this as **compound interest** since you are being paid interest on the interest as well as the principal. Compute the amount of money earned for a 20-year period from an initial investment of $10 000, compounded annually. Make a table to show the year, total interest, and total amount of money at the end of 20 years. Graph this new data of ordered pairs (years, total amount of money) on the same axes as in Exercise 1a. Use a different color for this new data. Compare the graphs of the two sets of data and explain how they are the same or different.

c. Suppose your bank decides to credit your account more frequently with the interest earned. Compounding semi-annually affects the rate. For example, a rate of 3% compounded semi-annually means that 3% ÷ 2 = 1.5% is the interest rate for each 6-month period. Each year, there are two such periods. Make charts to show the effects of

compounding semi-annually or every 6 months, for 10 years. Make graphs of the results on a new set of axes. Compare the results with those found in Exercise 1b. Write a conclusion about what you found.

d. COMPUTER ACTIVITY Complete the BASIC program below to compute and print results for any starting amount and interest rate. RUN it to check your results from Exercise 1b.

```
10 REM COMPOUND INTEREST
20 PRINT "Original amount:"
   :INPUT Principal
30 PRINT "Rate of interest:"
   :INPUT Rate
35 LET Percent = Rate/____
40 PRINT "Number of years?"
   :INPUT NumYrs
50 FOR Year = 1 TO _____
60 LET Earned In = Principal* _____
70 LET Principal = Principal + _____
80 PRINT Year,Earned In,Principal
90 NEXT Year
100 END
```

e. COMPUTER ACTIVITY Revise the program above to compute the semi-annual and quarterly interest for Exercise 1c.

f. COMPUTER ACTIVITY If your computer has a spreadsheet program,

What does x stand for? What does y stand for?

5. One of the most expensive purchases people make is an automobile. One of the greatest expenses in owning a car is the loss in value with each year's use. The **depreciation** is the difference between the original cost of the car and the amount received for it when it is traded or sold.

   a. A car that cost $16 000 was sold after three years for $6 400. What was the total depreciation? What was the average annual depreciation? What was the overall percentage rate of depreciation?

   b. The rate of depreciation is not the same each year. For a new car, it is the greatest during the first year. The chart gives percentages of what the car is worth that may apply to a certain class of cars.

NEW CAR DEPRECIATION	
END OF YEAR	% OF ORIGINAL COST
1	70
2	55
3	42
4	33
5	25

Study the depreciation of a $20 000 new car. Make a graph. Is this a growth or decay? Is it linear? When does the car seem to take its biggest drop in value? Explain why you think this happens.

c. Contact a bank, credit union, or library to obtain information on "blue book" values for used cars. Use these values to determine the depreciation percentages. Make a comparative study of different classes of cars. For example, imports versus American made, compacts versus full-size, cars versus trucks, or luxury versus economy. Use graphs to help show the rates of change. Which seem to depreciate the least? Why do you think this may be the case? Find the range of depreciation values in each year for different kinds of cars and explain how the depreciation rate might affect your decision when buying a car.

DOUBLE OR NOTHING?

Many real-world phenomena change by doubling, tripling, or halving. You will see how much light passes through a thickness of glass. You will see how mathematics is involved in photography. You will also investigate how a decay process occurs when a radioactive substance ages.

You will find how this kind of change is related to the size of the starting population, the growth or decay factor, and the number of intervals of growth or decay that takes place.

▶ GROUP PROJECT

Investigate growth and decay based on doubling and halving.

1. By folding a sheet of paper over and over, we can expand the thickness of the paper. There is a certain physical limit to the number of times a piece of paper, even the largest one you can think of, could be folded. This project investigates the mathematical theory of the folding of paper. Do you think that you could fold a sheet of paper enough times so that its thickness would be the same as the distance to the moon? How many folds would be needed?

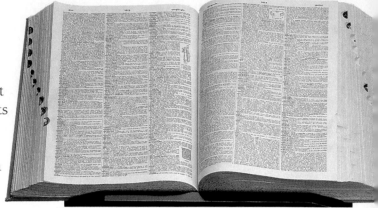

a. Take a sheet of paper and consider it a one sheet stack of paper. Fold it. Its thickness is doubled and now there are two sheets in your stack. Fold it again. How many sheets are there in your stack? Repeat the folding process and record the number of sheets with each fold. Make a chart and a graph to show how the number of sheets grows. How many sheets thick would your stack be if you made 10 folds? Use your calculator to fill in your chart to show the thickness after each folding.

b. Suppose the paper is 0.005 inch thick. Fill in your chart to show the thickness after each folding. How many folds are needed to produce at least one inch? Explain.

c. ![Portfolio Builder icon] **PORTFOLIO BUILDER** If the original thickness is t, then what is the thickness after one fold? Two folds? Three folds? n folds? Write an expression for n folds on page 118 of your Portfolio Builder.

d. Find the distance to the moon. How many folds would be needed to make a thickness that tall? Explain.

2. We often need to sort through large amounts of data to find the piece we are looking for. In this mind reading trick, you will notice how quickly successive halving can allow you to find what you are looking for. A mind reading trick can be explained as a decay process.

a. Have one group member secretly find and record a word in a dictionary. Have each group member predict how many questions you would need to ask in order to tell them their secret word.

b. Have another group member open the dictionary to where they would guess the middle to be. Read the first word at the top of the left-hand page.

c. Have the secret word writer say whether the word comes before or after the word at the top of the page. If the word comes before, use only the part of the dictionary from the first page to this middle page. If the word comes after, use only the part of the dictionary from the middle of the page to the last page.

d. Have another group member open the dictionary to find the middle of these new pages. Read the word at the top of the new left-hand page. Repeat the process, dividing the dictionary in half until the secret word is found.

e. Make a chart to show how many questions are needed until the word is found. Make a graph to show the decay process. Write an exponential expression to show the number of possible words.

f. Think about some other situation where you could use this divide-in-half method to perform a trick or explain a process.

▶ PARTNER PROJECT

Investigate changes that can be described with exponents.

Select and complete at least two of the following exercises.

3. Use cubes to investigate the following situation. Suppose you have a large cube of wood that is 1 unit on each edge.

   a. Suppose you cut the cube into smaller cubes, each $\frac{1}{2}$ unit on an edge. How many smaller cubes are there? These in turn are cut into cubes with edges half as long. What is their length? How many of these cubes result?

   b. Investigate the total surface area of the cubes after cutting. Make a table to organize this investigation. Include the number of cubes with each cut, the edge length of each cube, and the total surface area that results from each cut. Graph the results. What are the independent and dependent variables? Is the change in surface area linear or nonlinear? Explain. Write an exponential expression for the surface area.

   c. Suppose the original large cube is cut into smaller cubes, each with edges of $\frac{1}{3}$ unit. How many of these are

there? These are cut into smaller cubes with edges that are $\frac{1}{3}$ as long. How many cubes are there now, and what is the length of each edge?

   d. Investigate the total surface area of the cubes after cutting. Make a table and a graph, and compare your results with those in Exercise 3b.

4. Is there a photography class offered at your school? Have you ever used a single lens reflex camera to take pictures? If so, you know that changing the f-stop or the shutter speed can change the appearance of a picture dramatically.

   a. The camera aperture is an adjustable hole in a diaphragm covering the lens that controls the amount of light that will expose the film. On some cameras, you will see a ring around the lens that shows the size of this aperture measured in units called f-stops. Many cameras have the following f-stops: 1, 1.4, 2, 2.8, 4, 5.6, 8, 11, 16, and 22. Look at these numbers for patterns and explain the patterns you see.

   b. When beams of light go through a camera lens, they come together at a point, call the focal point. The **focal length** is the distance between this focal point and the center of the lens. Different camera lenses have different focal lengths. For example, a 35 mm camera typically has a lens of focal length 50 mm. Other common focal lengths for

camera lenses are 105 mm, 28 mm, and 15 mm. Use the equation,

$$\frac{\text{focal length}}{\text{f-stop number}} = \text{diameter to help}$$

you find the area of the aperture for each f-stop. You may need to refer back to Unit 6, Activity 4 where you found the area of a circle. Make a chart and a graph for two of the four focal lengths listed above. What is the shape of each graph?

c. How does decreasing the f-stop number by one stop affect the area of the aperture? Explain.

d. The shutter in the camera is like the lid of an eye when you blink. It opens and closes very quickly to expose the film to the light. Photographers can also control the shutter speed. A decrease in the f-stop number by one stop, halves the exposure time. If the proper exposure at stop 11 is a shutter speed of $\frac{1}{25}$ second, what would it be at each of the other stops? Make a new chart and graph. Explain what curve is represented.

5. Scientists use a unit called the **half life** to indicate the length of time it takes for an organic substance to disintegrate to half of its original value. Scientists use this unit of measure when working with radium, a substance that has a half life of 1 620 years. This means that half of the radioactive material in radium is gone after 1 620 years. After 2(1 620), or 3 240 years, only one-fourth of the original amount remains, and the substance has gone through two half-lives. Another way to look at this problem is with the exponential expression $\frac{1}{2}^0$, which represents 100% of the substance, $\frac{1}{2}^1$, representing 50% of the substance or one half-life, $\frac{1}{2}^2$ representing 25% of the substance or two half-lives, and so on.

a. You can simulate the idea of half life using number cubes or cubes with a mark on one face. Roll 100 such cubes. Remove the number cubes showing a 6, or the cubes showing the mark on a face, counting and

PORTFOLIO BUILDER

recording the number remaining. Repeat this until there are 50 or less remaining. The number of throws corresponds to the half life of the number cubes. Graph the results and describe the shape of the graph. Using theoretical probability, what fraction of the number cubes should remain each time? On page 118 of your Portfolio Builder, write an exponential expression to describe the half life of the number cubes.

b. **PORTFOLIO BUILDER** Change the number cube simulation by removing 1s and 6s, or cubes showing either one or two marks placed on faces. Collect and graph the data. Repeat the experiment. What is the half life for this 1 or 6 substance. Using theoretical probability, what fraction of the number cubes should remain each time? Write an exponential expression on page 118 of your Portfolio Builder to find the half-life for this 1 or 6 substance. Explain why and by how much the rate of decay changed.

c. Archaeologists are interested in a substance called carbon-14 because growing plants, and the animals that eat the plants, absorb carbon-14 during their lives. When living things die, the carbon-14 begins to disintegrate. By measuring the relative amount of carbon-14 remaining in a fossil, an estimate of the date when they died can be made. The half-life of carbon-14 is 5 700 years. It is used to help scientists find how long ago a plant or animal lived. What fraction of carbon-14 will remain in a sample after 57 000 years? Approximately, how old is a fossil bone where 1% of the initial amount of carbon-14 is found? Make a graph of the fraction of carbon-14 left as a function of time.

6. **PORTFOLIO BUILDER** The sun is a source of huge quantities of electromagnetic energy. Other sources include light bulbs, incandescent lamps, fluorescent lamps, and flames. What seems to fascinate us the most about the electromagnetic spectrum is visible light, which is less than 1% of the spectrum. In the passage of a beam of light through a medium, some of the energy is absorbed and transformed into heat, and some is scattered in all directions. When a beam of light strikes the surface between air and glass, some of the light is reflected back into the air and the remainder enters the glass. For each millimeter of thickness, a particular kind of glass allows 80% of the light to pass through. Make a chart and a graph to describe how much light, l, passes through n millimeters of thickness of this glass. Then, on page 118 of your Portfolio Builder, write an expression to describe this situation.

7. The Richter magnitude scale is the scale most commonly used to measure the strength of an earthquake. It is also most often misunderstood.

   a. Each whole number step of magnitude on the scale represents an increase of 10 times in the intensity. Suppose that an earthquake of magnitude 4.0 has an intensity of 1 unit. Find the relative intensities of earthquakes with magnitudes of 2, 3, 5, 6, 7, 7.1(Loma Prieta, California, 1989), 7.3(Japan, 1948), 8.0(San Francisco, 1906), and 9.2(Alaska, 1964). Graph the ordered pairs (magnitude, relative intensity). What is the shape of this graph?

   b. Each whole number step of magnitude on the scale also represents an increase of 31 times the amount of energy released. If the earthquake of magnitude 4.0 has e units of energy released, find the relative energy released at the other quake magnitudes.

   c. Compare the relative magnitudes and amounts of energy released between the San Francisco and Loma Prieta earthquakes. Write an article for a newspaper that explains this comparison and also elearly explains the Richter Scale.

POPULATION EXPLOSIONS!

In this unit, you have seen how growth and decay can occur. One of the most interesting and important kinds of growth occurs with populations, such as bacteria, deer, or people.

In this activity, you will study how the size of populations can change, sometimes dramatically! How does the growth of a city typically occur? How has your city changed in size over the past century? You will see how the rapid changes, such as doubling or halving, affect population size. Once again, exponents can be used in interpreting such change.

▶ GROUP PROJECT

Explore how populations can grow exponentially.

You have seen how living things grow and eventually decay. Even populations grow or perhaps decline.

1. Microscopic populations often grow very rapidly. To investigate the growth of a population of mold, as a class, use this procedure:

♦ With rubber cement, fasten graph paper to the bottom inside of a 9-inch aluminum pie tin. With the origin near the center, draw the x- and y-axes with a lead pencil or waterproof ink. This will make about 10 tins for the class.

♦ Mix an envelope of colorless gelatin with 2 tablespoons of cold water. Mix a beef bouillon cube with 2 tablespoons of very hot water. After each is dissolved, combine the two mixtures. This should be enough for 10-12 tins.

♦ The gelatin mix will set as it cools. Therefore, immediately pour just enough into each tin to cover the graph paper with a thin layer. While it is setting, it will become contaminated by mold spores naturally floating in the air.

♦ After 5 minutes, cover the tin with enough transparent clinging wrap to fold down on all sides and to seal with tape or a rubber band.

♦ Store in a dark place, such as a drawer or cupboard, at a fairly constant and warm temperature.

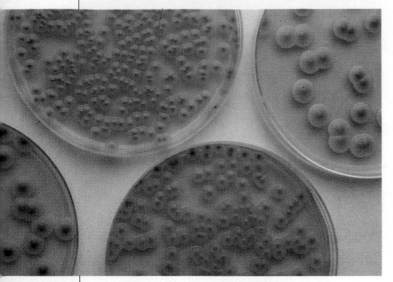

a. Every school day at the same time, observe and record the total number of graph paper squares covered by growing mold. Collect data for at least three weeks. Take a photograph of your mold.

b. For each day, try to predict the next value. Compute the increase in area for each day. Compute the percent of increase in area for each day. What is the rate of growth, and how does it seem to change?

c. Graph the data. Use the graph to help interpolate the missing values for the weekends. What is the shape of the graph?

d. You should see an S-curve typical of the growth of populations. In the early life of a population, there is slow growth, referred to as the **lag phase**. Then the rapid growth, known as **exponential growth**, is followed by slower growth leading to a **stationary**

phase. Mark these on your graph. How long was the lag phase? The exponential growth? The approach to a stationary phase? Did your mold population reach a stationary phase, or even begin to decline? Explain.

e. Why do you think the mold population grew? What factors promoted the growth? Limited the growth? How might the growth of mold population be related to other population growth? Explain.

2. The population of California for the last 100 years is presented below.

CALIFORNIA POPULATION

YEAR	POPULATION
1890	1 213 398
1900	1 485 053
1910	2 377 549
1920	3 426 861
1930	5 677 251
1940	6 907 387
1950	10 586 223
1960	15 717 204
1970	19 953 134
1980	23 667 565
1990	29 760 021

a. Graph the data to show how the population of California has changed. What is the shape of the curve? Identify the three phases of growth and explain why it does or does not fit the S-curve pattern. What was happening in the history of California that could account for these rapid changes? Explain.

b. To find out more about the growth rate, compute and record the ratios that compare population sizes every 10 years. Change these ratios to decimals and percents. Use these to determine the least and greatest change, the range of the changes in growth, the median change, and the mean change.

c. Predict the population in the years 2000 and 2020. Compare your value with a U.S. Bureau of the Census projection of 33 963 000 in the year 2000. Explain the difference in your prediction and the Bureau of Census prediction.

d. Use your graph and estimate the population in 1945.

e. Assume the growth was the same rate for each year from 1940 to 1950. This means that the growth from 1940 to 1945 will equal the growth from 1945 to 1950. For a quantity that is growing exponentially, the method below can be used to find the amount in the middle of an interval of growth. If there is an amount a at the beginning of a period of growth, and b at the end, then $\frac{a}{x} = \frac{x}{b}$ so $x = \sqrt{ab}$.

This value is called the **geometric mean**. Use it to estimate the populations of California in the middle of each interval, such as 1895, 1905, and 1915. Check these estimates to see how well they fit your graph. How close were they? Explain.

f. The growth curve for California does not appear to be approaching a stationary phase. Why not? What might cause the growth to slow down? Explain.

▶ HOMEWORK PROJECT

3. Use the chart below to make a graph for Earth's population since 1500. Is the growth linear or nonlinear? Explain.

EARTH'S POPULATION

DATE	POPULATION (MILLIONS)
1500	460
1600	579
1700	679
1750	770
1800	954
1900	1 633
1920	1 862
1940	2 295
1960	3 019
1980	4 450
2000 (est)	6 251
2020 (est)	8 281

a. Use your graph to predict today's population of Earth. Compare the current value with your prediction. Explain any differences.

b. One way to express the rate of change is to tell how long it is before it doubles. Use your graph to predict the current length of time for Earth's population to double. If the doubling time is shorter today, what does this indicate about the rate of population growth? Explain.

c. According to United Nations information, the population of Earth was about 4 billion in 1975, and at that time it was predicted to double in 35 years. Study the doubling period. How has it changed? Find how long it took for the population to double in 1500, 1600, 1700, 1800, and 1900. Use your chart and the geometric mean to estimate the number of years until the population doubles. Graph the approximate ordered pairs, (century, time to double). What is the shape of the graph? Explain.

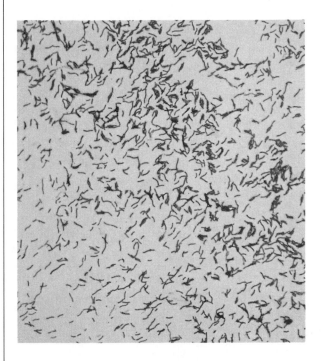

4. **PORTFOLIO BUILDER** Bacteria are very tiny, single-celled organisms that are thought to be the most ancient forms of life on Earth. They live in huge numbers in almost every habitat, including your body. Some are harmful and can cause disease. Food can spoil rapidly because of bacteria. Bacteria is reproduced by splitting itself in two. In good conditions, some bacteria can do this every 20 minutes, so it soon becomes surrounded by enormous numbers of its offspring. Find the number of bacteria produced from one initial cell in a 6-hour period. On page 118 of your Portfolio Builder, write an equation you can use to find this number.

5. The population of Manhattan Island in New York City, rounded to the nearest thousand, is given below.

MANHATTAN ISLAND POPULATION

YEAR	POPULATION
1790	33 000
1800	61 000
1810	96 000
1820	124 000
1830	203 000
1840	313 000
1850	516 000
1860	814 000
1870	942 000
1880	1 165 000
1890	1 441 000
1900	1 850 000
1910	2 332 000
1920	2 284 000
1930	1 867 000
1940	1 890 000
1950	1 960 000
1960	1 698 000
1970	1 539 000
1980	1 428 000
1990	1 488 000

a. Graph the data. Identify the shape of the curve and use ratios to determine the rate of change (number of people per year) in each decade.

b. Record the percent of population change from one decade to the next. Which decade had the largest rate of change? Describe whether the graph shows growth or decay.

c. In what years did Manhattan Island lose population? Determine the average decrease per year. What factors might explain this change?

d. Examine population change in other cities in the United States by selecting some city of interest to you. Obtain population data over several decades. Make a graph of the data. Explain how the size of the city has changed and why these changes may have occurred. Make a list of factors that can increase the size of a city and those that can cause the population to decrease. Based on the data, make projections for the next few decades.

CHECKING YOUR PROGRESS

This assessment will involve your knowledge of the mathematics you have learned in this unit. In Part I, you will work with a partner to solve some situations that use ideas from this unit. In Part II, you will work in a group to apply what you have learned in this unit. In Part III, you will complete and submit your work on the Unit Problem.

PART I PARTNER ASSESSMENT

With your partner, solve the following.

1. Suppose that a woman receives her weekly salary of $400 every Friday.
   a. If she spends half of her money each day, how much will she have left on the following Friday? Make a graph to show the current funds for each day of the week. Explain the shape of the graph and relate it to growth or decay situations.
   b. Suppose her employer decides to pay her $800 every other Friday. If she has the same spending habit, will she be in better or worse financial condition on the next payday? Make a graph to compare the two situations. Explain your answer.

c. Suppose the woman decides to spend at a slower rate, but follows a similar spending pattern. She wants to have at least $25 left on the next payday. What fraction should she spend each day if she is paid $400 weekly? $800 bi-weekly? Make graphs to compare these approaches.

PART II GROUP ASSESSMENT

2. You can make interesting growth patterns by coloring graph paper.
   a. On graph paper, color a square. With a different color or pattern, shade all of the squares that touch it, including corners. With a new color or pattern, shade all of the squares that touch these. Repeat this. For each color, record the pattern number, the number of new squares, and the total number of squares in a chart. Search for number patterns in your chart. Explain these patterns.

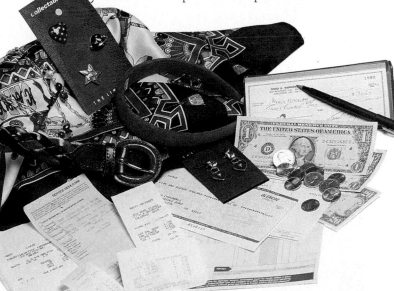

b. Graph the ordered pairs (number of new colors, number of new squares). Write an expression for the number of squares added. Describe the shape of your graph. On a separate axes, graph the ordered pairs (number of new colors, total number of squares). Describe the shape of this graph.

c. How does the area of each successive square grow?

d. Select and color two adjacent squares. With a different color, shade all touching squares. Repeat this again and again. Make a chart and graph as you did before. How does the area of the shape grow? Write an expression for the growth.

e. Select and color three squares in an "L" shape. As before, investigate how the area grows with each successive coloring of touching squares.

f. Make up four starting shapes, each more complex than the ones above. As before, investigate how the area grows when you color touching squares with successively different colors or patterns.

g. What can you conclude about how the areas of shapes grow in the previous problem? Explain.

3. Pick one of the following topics for your group to research. Collect data and articles about your topic to use in a written report.

◆ Population growth is at the root of worldwide concern over our ability to provide sufficient food, shelter, and "space" for all people. Research a country, such as Bangladesh, where the food and shelter cannot support the population. What are the causes for the population growth, the food shortages, and the lack of shelter?

◆ Large epidemics affect population growth. Research an epidemic such as the Black Plague from 1347-1351. What were the causes of this plague? How was the population growth affected? Several years ago, a newspaper article compared the AIDS crisis with the Black Plague. How is AIDS affecting the world population growth? How will it affect the growth rate of cities? How does this rate compare to the Black Plague's?

◆ Research an animal that is extinct or is becoming extinct such as the carrier pigeon, the condor, or the bald eagle.

◆ Explain the factors that contributed to the animal's extinction.

PART III UNIT PROBLEM

Complete your cooperative and individual work on your unit problem. Write and present your report.

ENDING UNIT 8

Now that you have completed Unit 8, refer to page 119 of your Portfolio Builder. You will find directions for building your personal portfolio from items you completed in this unit. Then, on page 120 of your Portfolio Builder, you will find directions for completing your portfolio for this school year.

REFERENCE SECTION

REFERENCE SECTION

They're good for you, but many still shun fruit, veggies

By Nanci Hellmich
USA TODAY

We still aren't eating enough fruits and vegetables. In fact, the average person falls short by 500 servings a year, suggests a survey out Wednesday.

It reiterates what other studies have shown — and what mothers have known for years.

"For many cancers, persons with high fruit and vegetable intake have about half the risk of people with low intakes." says Dr. Peter Greenwald, National Cancer Institute.

The telephone survey of 2,837 people over 18 says:

▶*77% don't eat the recommended five or more servings of fruits and vegetables a day.

▶*People eat an average of 3 1/2 servings a day.

▶*Women eat about four servings a day, compared to men, who eat about three.

▶*Only 8% of adults think they should eat five or more servings of fruits and vegetables each day; 66% think two or fewer are sufficient.

The poll is part of the "5 a Day for Better Health" campaign sponsored by NCI and Produce for Better Health, a produce industry group. The program was announced last fall; many consumers have seen brochures in their grocery stores. Other survey findings:

▶*People over 65 eat about four servings of fruits and vegetables a day; people 18 to 34, about three.

How much is enough

If you're trying to eat five or more servings of fruits and vegetables a day, take note: Lettuce on your hamburger or mushrooms on your pizza don't count for a serving.

The U.S. Department of Agriculture's Dietary Guidelines recommend two to four servings of fruits a day; three to five servings of vegetables.

How to count your servings:

▶*Fruit. 1 medium piece fresh; 1/2 cup of cooked or raw fruit; 3/4 cup juice.

▶*Vegetables. 1 cup raw leafy greens; 1/2 cup others.

You don't have to eat just raw radishes or celery sticks to get more veggies in your diet, says Felicia Busch, American Dietetic Association.

Her suggestions: Add vegetables to casserole. Top cereals with fruit — try dates, dried apricots, papaya.

▶*41% of all people believe it's likely that eating fruits and vegetables helps prevent cancer; 52% believe it prevents heart disease; 61% believe it helps in losing or maintaining weight.

▶*About 25% always or usually eat vegetables cooked in some type of fat; about 20% always or usually add butter, cream or cheese sauce.

"It's not that fruits and vegetables are unattractive foods," says Felicia Busch, a spokeswoman for the American Dietetic Association. "It's just not what pops into many people's minds when they're hungry."

A smoker at home harms kids

Study: Children exposed to cigarettes have higher health risk

The Associated Press

WASHINGTON — Children who live in households with smokers are much more likely to be in fair or poor health than are children never exposed to cigarette smoke, a government study reported today.

"I can't think of a more compelling reason for parents to quit smoking than ensuring their children's chance for a healthy life," Health and Human Services Secretary Louis Sullivan said in releasing the report.

The study by the National Centers for Health Statistics found that 4.1 percent of children in households with current smokers were in fair to poor health, compared with 2.4 percent of children never exposed to tobacco smoke.

About half of all children in the United States who are 5 years old or younger have been exposed to cigarette smoke, according to the study. More than one-quarter of all children in this age group were exposed to smoke both before and after birth.

The study was based on a 1988 smoking survey that included a question about the health status of children 5 years of age and younger in the household. The survey sample included 5,356 children in that age group, and the child's health was reported by the

Please see KIDS, A5•>

Kids: Smoker in home can hurt health of children, study finds

► Continued from A1

household respondent, usually a parent.

The report said that while children's health appears to be associated with various exposures to cigarette smoking, the results should be interpreted "with caution" because they do not take into account possible variations in sampling and perceived health status.

Children in families with lower incomes and less education were more likely to have been exposed to cigarette smoking and were more likely to have been reported to be in fair or poor health, the report said.

In families with incomes of less than $10,000, about two-thirds of young children were exposed to smoke, compared with about one-third of children in families where the income was $40,000 or more.

The ratio was about the same for families where the mother had not completed high school compared with those where the mother had one or more years of college.

Children in families with lower incomes and less education were more likely to have been exposed to cigarette smoking and were more likely to have been reported to be in fair or poor health, the report said.

About 60 percent of black children had been exposed to smoke compared with 49 percent of white children.

Hispanic children were less likely than non-Hispanic children to have been exposed. While 51 percent of non-Hispanic children had ever been exposed, 44 percent of Hispanic children had been, and the proportion dropped to 40 percent for Mexican-American children.

	Population Estimate mid - 1992 (millions)	Birth Rate (per 1 000 pop.)	Population "Doubling Time" in Years (at current rate)	Population Projected to 2010 (millions)	Infant Mortality Rate (per 1 000 live births)	Urban Population (%)	Per Capita GNP, 1990 (US$)
AFRICA	**654**	**43**	**23**	**1,085**	**99**	**30**	**630**
NORTHERN AFRICA	**147**	**35**	**27**	**216**	**72**	**43**	**1 070**
Algeria	26.0	35	28	37.9	61	50	2 060
Egypt	55.7	32	28	81.3	73	45	600
Libya	4.5	37	23	7.1	64	76	—
Morocco	26.2	33	29	36.0	73	46	950
Sudan	26.5	45	22	42.2	87	20	—
Tunisia	8.4	27	33	11.3	44	53	1 420
Western Sahara	0.2	49	25	0.3	—	—	—
WESTERN AFRICA	**182**	**47**	**23**	**312**	**111**	**23**	**410**
Benin	5.0	49	23	8.9	88	39	360
Burkina Faso	9.6	50	21	17.0	121	18	330
Cape Verde	0.4	41	21	0.7	41	33	890
Côte d'Ivoire	13.0	50	19	25.5	92	43	730
Gambia	0.9	46	27	1.6	138	22	260
Ghana	16.0	44	22	26.9	86	32	390
Guinea	7.8	47	28	11.6	148	22	480
Guinea — Bissau	1.0	43	35	1.5	151	27	180
Liberia	2.8	47	22	5.5	144	44	—
Mali	8.5	52	23	14.2	113	22	270
Mauritania	2.1	46	25	3.5	122	41	500
Niger	8.3	52	22	15.4	124	15	310
Nigeria	90.1	46	23	152.2	114	16	370
Senegal	7.9	45	25	13.1	84	37	710
Sierra Leone	4.4	48	27	7.3	147	30	240
Togo	3.8	50	19	7.1	99	24	410
EASTERN AFRICA	**206**	**47**	**22**	**359**	**110**	**19**	**230**
Burundi	5.8	47	21	10.1	111	5	210
Comoros	0.5	48	20	0.9	89	26	480
Djibouti	0.4	46	24	0.7	117	79	—
Ethiopia	54.3	47	25	94.0	139	12	120
Kenya	26.2	45	19	44.8	62	22	370
Madagascar	11.9	45	22	21.3	115	23	230
Malawi	8.7	53	20	14.9	137	15	200
Mauritius	1.1	21	48	1.3	20.4	41	2 250
Mozambique	16.6	45	26	26.6	136	23	80
Reunion	0.6	24	38	0.8	13	62	—
Rwanda	7.7	51	20	14.4	117	7	310
Seychelles	0.1	24	44	0.1	13	52	4 670
Somalia	8.3	49	24	13.9	127	24	150
Tanzania	27.4	50	20	50.2	105	21	120
Uganda	17.5	52	19	32.5	96	10	220
Zambia	8.4	51	18	15.5	76	49	420
Zimbabwe	10.3	41	22	17.0	61	26	640

Continued on next page

REFERENCE SECTION

POPULATION STATISTICS, AFRICA

	Population Estimate mid - 1992 (millions)	Birth Rate (per 1,000 pop.)	Population "Doubling Time" in Years (at current rate)	Population Projected to 2010 (millions)	Infant Mortality Rate (per 1000 live births)	Urban Population (%)	Per Capita GNP, 1990 (US$)
MIDDLE AFRICA	**72**	**45**	**23**	**122**	**97**	**38**	**460**
Angola	8.9	47	25	14.9	132	26	—
Cameroon	12.7	44	22	23.1	85	42	940
Central African Republic	3.2	44	27	4.9	141	43	390
Chad	5.2	44	28	7.7	127	30	190
Congo	2.4	43	24	3.9	114	41	1 010
Equatorial Guinea	0.4	43	26	0.6	112	28	330
Gabon	1.1	41	28	1.4	99	43	3 220
Sao Tome and Principe	0.1	35	28	0.2	71.9	38	380
Zaire	37.9	46	22	65.6	83	40	230
SOUTHERN AFRICA	**47**	**35**	**26**	**76**	**57**	**52**	**2 390**
Botswana	1.4	40	23	2.4	45	24	2 040
Lesotho	1.9	41	24	3.1	95	19	470
Namibia	1.5	43	22	2.9	102	27	—
South Africa	41.7	34	26	66.0	52	56	2 520
Swaziland	0.8	44	22	1.5	101	23	820

Source: 1992 World Population Data Sheet

Reference Section

THE PRESIDENTS OF THE UNITED STATES

PRESIDENT	BORN	AGE AT INAUGURATION	SERVED	DIED	AGE AT DEATH
1. George Washington	Feb. 22, 1732	57	1789 - 1797	Dec. 14, 1799	67
2. John Adams	Oct. 30, 1735	61	1797 - 1801	July 4, 1826	90
3. Thomas Jefferson	Apr. 13, 1743	57	1801 - 1809	July 4, 1826	83
4. James Madison	Mar. 16, 1751	57	1809 - 1817	June 28, 1836	85
5. James Monroe	Apr. 28, 1758	58	1817 - 1825	July 4, 1831	73
6. John Quincy Adams	July 11, 1767	57	1825 - 1829	Feb. 23, 1848	80
7. Andrew Jackson	Mar. 15, 1767	61	1829 - 1837	June 8, 1845	78
8. Martin Van Buren	Dec. 5, 1782	54	1837 - 1841	July 24, 1862	79
9. William H. Harrison	Feb. 9, 1773	68	1841	Apr. 4, 1841	68
10. John Tyler	Mar. 29, 1790	51	1841 - 1845	Jan. 18, 1862	71
11. James K. Polk	Nov. 2, 1795	49	1845 - 1849	June 15, 1849	53
12. Zachary Taylor	Nov. 24, 1784	64	1849 - 1850	July 9, 1850	65
13. Millard Fillmore	Jan. 7, 1800	50	1850 - 1853	Mar. 8, 1874	74
14. Franklin Pierce	Nov. 23, 1804	48	1853 - 1857	Oct. 8, 1869	64
15. James Buchanan	Apr. 23, 1791	65	1857 - 1861	June 1, 1868	77
16. Abraham Lincoln	Feb. 12, 1809	52	1861 - 1865	Apr. 15, 1865	56
17. Andrew Johnson	Dec. 29, 1808	56	1865 - 1869	July 31, 1875	66
18. Ulysses S. Grant	Apr. 27 1822	46	1869 - 1877	July 23, 1885	63
19. Rutherford B. Hayes	Oct. 4, 1822	54	1877 - 1881	Jan. 17, 1893	70
20. James A. Garfield	Nov. 19, 1831	49	1881	Sept. 19, 1881	49
21. Chester A. Arthur	Oct. 5, 1829	51	1881 - 1885	Nov. 18, 1886	57
22. Grover Cleveland	Mar. 18, 1837	47	1885 - 1889	June 24, 1908	71
23. Benjamin Harrison	Aug. 20, 1833	55	1889 - 1893	Mar. 13, 1901	67
24. Grover Cleveland	Mar. 18, 1837	55	1893 - 1897	June 24, 1908	71
25. William McKinley	Jan. 29, 1843	54	1897 - 1901	Sept. 14, 1901	58
26. Theodore Roosevelt	Oct. 27, 1858	42	1901 - 1909	Jan. 6, 1919	60
27. William H. Taft	Sept. 15, 1857	51	1909 - 1913	Mar. 8, 1930	72
28. Woodrow Wilson	Dec. 29, 1856	56	1913 - 1921	Feb. 3, 1924	67
29. Warren G. Harding	Nov. 2, 1865	55	1921 - 1923	Aug. 2, 1923	57
30. Calvin Coolidge	July 4, 1872	51	1923 - 1929	Jan. 5, 1933	60
31. Herbert C. Hoover	Aug. 10, 1874	54	1929 - 1933	Oct. 20, 1964	90
32. Franklin D. Roosevelt	Jan. 30, 1882	51	1933 - 1945	Apr. 12, 1945	63
33. Harry S. Truman	May 8, 1884	60	1945 - 1953	Dec. 26, 1972	88
34. Dwight D. Eisenhower	Oct. 14, 1890	62	1953 - 1961	Mar. 28, 1969	78
35. John F. Kennedy	May 29, 1917	43	1961 - 1963	Nov. 22, 1963	46
36. Lyndon B. Johnson	Aug. 27, 1908	55	1963 - 1969	Jan. 22, 1973	64
37. Richard M. Nixon	Jan. 9, 1913	56	1969 - 1974		
38. Gerald R. Ford	July 14, 1913	61	1974 - 1977		
39. James E. Carter, Jr.	Oct.1, 1924	52	1977 - 1981		
40. Ronald W. Reagan	Feb. 6, 1911	69	1981 - 1989		
41. George H.W. Bush	June 12, 1924	64	1989 - 1993		
42. William J. Clinton	Aug. 19, 1946	46	1993 -		

PLANETS OF OUR SOLAR SYSTEM

PLANETS/ SUN	DIAMETER	DISTANCE FROM SUN	NUMBER OF MOONS	1 ROTATION*	ORBIT
Mercury	3 100 miles	36 million miles	0	59 days	88 days
Venus	7 500 miles	67 million miles	0	243 days	225 days
Earth	7 926 miles	93 million miles	1	24 hours	365 days
Mars	4 218 miles	142 million miles	2	24.4 hours	687 days
Jupiter	89 400 miles	483 million miles	16	10 hours	11.86 years
Saturn	75 000 miles	886 million miles	20	10.4 hours	29.46 years
Uranus	32 300 miles	1.8 billion miles	15	17 hours	84 years
Neptune	30 000 miles	2.8 billion miles	3	18 - 22 hours	165 years
Pluto	1 900 miles	3.7 billion miles	1	6.4 days	248 years
Sun	870 000 miles				

*hours, days, and years are Earth time

Reference Section

POPULATION, SELECTED STATES AND CITIES

STATE AND CITY	POPULATION (1990)	AREA (SQUARE MILES)
Arizona		
Glendale	148 000	52.2
Mesa	288 000	108.6
Phoenix	983 000	419.9
Scottsdale	130 000	184.4
Tempe	142 000	39.5
Tucson	405 000	156.3
California		
Anaheim	266 000	44.3
Chula Vista	135 000	29.0
Fresno	354 000	99.1
Inglewood	110 000	9.2
Los Angeles	3 485 000	469.3
Oakland	372 000	56.1
Pomona	132 000	22.8
Sacramento	369 000	96.3
San Diego	1 111 000	324.0
San Francisco	724 000	46.7
San Jose	782 000	171.3
Santa Rosa	113 000	33.7
Stockton	211 000	52.6
Connecticut		
Bridgeport	142 000	16.0
Hartford	140 000	17.3
New Haven	130 000	18.9
Stamford	108 000	37.7
Waterbury	109 000	28.6
Florida		
Ft. Lauderdale	149 000	31.4
Hialeah	188 000	19.2
Jacksonville	635 000	758.7
Miami	359 000	35.6
Orlando	165 000	67.3
St. Petersburg	239 000	59.2

POPULATION, SELECTED STATES AND CITIES

STATE AND CITY	POPULATION (1990)	AREA (SQUARE MILES)
Pennsylvania		
Allentown	105 000	17.7
Erie	109 000	22.0
Philadelphia	1 586 000	135.1
Pittsburgh	370 000	55.6
Tennessee		
Chattanooga	152 000	118.4
Knoxville	165 000	77.3
Memphis	610 000	256.0
Nashville-Davidson	448 000	473.3
Texas		
Abilene	107 000	103.1
Amarillo	158 000	87.9
Arlington	262 000	93.0
Austin	466 000	217.8
Dallas	1 007 000	342.4
El Paso	515 000	245.4
Houston	1 631 000	539.9
San Antonio	936 000	333.0
Virginia		
Arlington	171 000	25.9
Chesapeake	152 000	340.7
Hampton	134 000	51.8
Norfolk	261 000	53.8
Richmond	203 000	60.1
Washington		
Seattle	516 000	83.9
Spokane	177 000	55.9
Tacoma	177 000	48.1

TABLE OF WEIGHTS AND MEASURES

Customary Conversions

Length	1 foot (ft)	=	12 inches (in.)
	1 yard (yd)	=	3 feet
	1 mile (mi)	=	5 280 feet
Weight	1 pound (lb)	=	16 ounces (oz)
	1 ton (T)	=	2 000 pounds
Capacity	1 cup (c)	=	8 fliud ounces (fl oz)
	1 pint (pt)	=	2 cups
	1 quart (qt)	=	2 pints
	1 gallon (gal)	=	4 quarts

Metric Conversions

Length	1 centimeter (cm)	=	10 millimeters (mm)
	1 meter (m)	=	100 centimeters
	1 kilometer (km)	=	1 000 meters
Mass	1 gram (g)	=	1 000 milligrams (mg)
	1 kilogram (kg)	=	1 000 grams
Capacity	1 liter (L)	=	1 000 milliliters (mL)
	1 kiloliter (kL)	=	1 000 liters

Changing Units in the Metric System

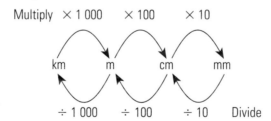

Multiply $\times$ 1 000 $\times$ 100 $\times$ 10

km   m   cm   mm

$\div$ 1 000 $\div$ 100 $\div$ 10   Divide

Time Conversions

1 minute (min)	=	60 seconds (s)
1 hour (h)	=	60 minutes
1 day (d)	=	24 hours
1 year (yr)	=	356 days

Square

s = length of side

Area = s^2

Rectangle

l = length

w = width

Area = $l\,w$

Parallelogram

b = base

h = height

Area = $b\,h$

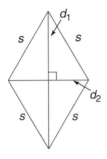

Rhombus

s = side

d_1 = diagonal 1

d_2 = diagonal 2

Area = $\frac{1}{2}d_1 d_2$

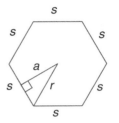

Regular Hexagon

Perimeter = $6s$

r = radius

a = apothem

Area = $\frac{1}{2}aP$ or

Area = $6\left(\frac{s^2\sqrt{3}}{4}\right)$

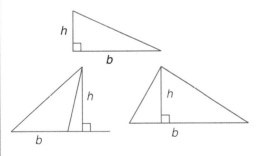

Triangle

b = base

h = height

Area = $\frac{1}{2}bh$

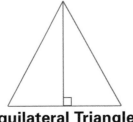

Equilateral Triangle

s = side

Area = $\frac{s^2\sqrt{3}}{4}$

Circle

r = radius

d = diameter

$d = 2r$

C = circumference

$C = 2\pi r$

$C = 2d$

Area = $\pi\,r^2$

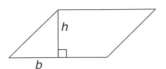

Cube
s = length of edge
Volume = s^3
Surface Area = $6\,s^2$

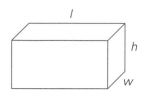

Rectangular Prism
h = height
w = width
l = length
Volume = $l\,w\,h$

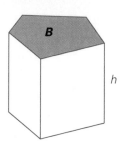

Right Prism
h = height
B = Area of Base
Volume = $B\,h$

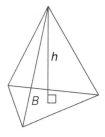

Right Pyramid
h = height
B = Area of Base
Volume = $\frac{1}{3}\,B h$

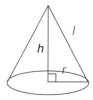

Right Circular Cone
r = radius
h = height
l = slant height
Volume = $\frac{1}{3}\,\pi r^2 h$

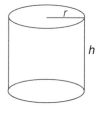

Right Circular Cylinder
r = radius
h = height
Volume = $\pi r^2 h$

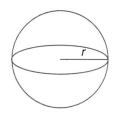

Sphere
r = radius
Volume = $\frac{4}{3}\,\pi r^3$

REFERENCE SECTION

CONCEPTS AND APPLICATIONS

The Concepts and Applications section contains key mathematical concepts presented through completely worked-out examples and applications. The topics in this section allow you to review important mathematical skills that you need to apply as you complete an activity. As you are working an activity, refer to this section if you feel that you need a quick review.

NUMBER LINES

- A **number line** is a line with a numerical scale. A **scale** includes all possible values in a given set of data. It does not necessarily show the points for 0 or 1. The **interval** is the measure of the distance between consecutive values labeled on the scale. The **range** for a set of data is the difference between the greatest number and the least number in the data.

EXAMPLE A Construct a basic number line.

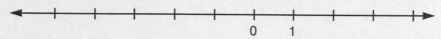

Start with a line or line segment. Label some point to be 0. The number 1 is then assigned to some point. The **unit distance** is the distance from 0 to 1. Use this distance to mark the points for other whole numbers, such as 2, 3, 4, and so on.

    Think Where would the point for -1 be found on the number line above? $\frac{1}{2}$? 0.75? -2 $\frac{3}{4}$?

You often need to make a number line to show a particular scale.

EXAMPLE B Construct a number line for a scale from 45 to 75 with an interval of 5.

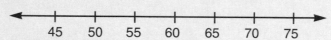

    Think Is 55 included on this scale? 69? 125? 0? Why?

Sometimes it is necessary to fit a scale onto a segment having a given length.

APPLICATION: Teens and Telephones

How much time does the average teenager talk on the phone? Jordan took a survey of his classmates to see how long they talked on the phone, in hours, during the past 7 days.

Paula	15.5	Ashley	18	Olivia	10.5	Emilio	12
Nick	12.5	Amber	11.5	Marcus	13	Ramon	11.5
Marni	16	Matt	14	Carolyn	13	Benito	15

Make a number line that is 12 cm long to show this data.

Find the range: $18 - 10.5 = 7.5$

The scale can be 10 to 18 with an interval of 1. You must fit eight intervals on a scale 12 cm long. Thus, the actual size of the interval is $12 \div 8$ or 1.5 cm.

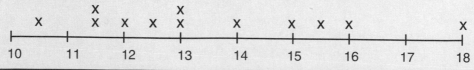

VENN DIAGRAMS

- A **Venn diagram** is a figure using circles in a rectangle to show relationships among sets of items. The **universal set** includes everything inside the rectangle. A set of elements is shown within a circle. A **subset** would be a set included in the original set. If the circles overlap, an **intersection set** is found where the circles overlap. The **union** of two sets would include everything inside both circles. The **complement** of a set would include everything not in that set.

EXAMPLE If set A = {1, 3, 4, 5, 7}, set B = {2, 5, 6, 7, 10} and the universal set U = {1, 2, 3, 4, 5, 6, 7, 8, 9, 10}, construct a Venn diagram to show the relationship among the sets.

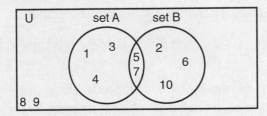

Think What is the intersection of sets A and B? the complement of set A? the union of sets A and B? a subset of set B?

APPLICATION: Pizza Preferences
José asked 42 people which pizza topping they liked, pepperoni (P), sausage (S), both, or neither. Use the Venn diagram to answer the questions.

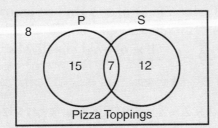

Pizza Toppings

How many liked pepperoni?	15 + 7 = 22
How many liked only sausage?	12
How many liked both pepperoni and sausage?	7
How many liked pepperoni or sausage, but not both?	15 + 12 = 27
How many did not like either pepperoni or sausage?	42 − (15 + 7 + 12) = 8

CONCEPTS AND APPLICATIONS

NUMBERS AND NUMERALS

- A *fraction* is a number for a part of a whole, for a measure, for a ratio, or for a quotient. A fraction has a numerator and a denominator.

EXAMPLE A   **Describe models or situations that use a fraction.**

   a.   **Part of a whole**   If a bag of apples has 3 red and 5 yellow, then $\frac{3}{8}$ (three-eighths) of the set of apples are red and $\frac{5}{8}$ (five-eighths) are yellow.

   b.   **A measure**   When a pizza is cut into six equal-sized pieces and two are eaten, then $\frac{4}{6}$ of the pizza remains.

   c.   **A ratio**   If a car can be driven 30 miles using 1 gallon of gasoline, then $\frac{60}{2}$ is a fraction for the same rate.

   d.   **A quotient**   To share three same-sized pizzas equally among four persons, each person would eat $3 \div 4$ or $\frac{3}{4}$ of a pizza.

   Think   What situation from your life can be described by the fraction $\frac{2}{3}$?

- A *proper fraction* has a numerator less than its denominator. An *improper fraction* has a numerator equal to or greater than its denominator. A *mixed number* has a whole number and a proper fraction. The *reciprocal* of a fraction has the numerator and denominator interchanged. The product of the number and its reciprocal is 1.

EXAMPLE B   **Name the reciprocal of $\frac{2}{3}$.**

   The reciprocal is $\frac{3}{2}$ since $\frac{2}{3} \times \frac{3}{2} = 1$.

   Think   Does every number have a reciprocal? When are a number and its reciprocal equal?

- **Decimals** are another way to write fractions when the denominators are 10, 100, 1 000, and so on. The place values of decimals are tenths, hundredths, thousandths, and so on.

EXAMPLE C Write expanded numerals to show place values for each number.

  a. 1.34 $1.34 = 1 + \frac{3}{10} + \frac{4}{100}$

  b. 0.123 $0.123 = \frac{1}{10} + \frac{2}{100} + \frac{3}{1\,000}$

  c. 0.00502 $0.00502 = \frac{5}{1\,000} + \frac{2}{100\,000}$

  Think Order from least to greatest: 0.1, 0.02, 0.003, 0.0004.

- The **square root** of a number is one of the two equal factors of the number. The symbol $\sqrt{}$, called a **radical sign**, is used to indicate a nonnegative square root.

EXAMPLE D Find each value.

  a. $\sqrt{9}$ $\sqrt{9} = 3$, since $3 \times 3 = 9$

  b. $\sqrt{121}$ $\sqrt{121} = 11$, since $11 \times 11 = 121$

  c. $\sqrt{\frac{1}{4}}$ $\sqrt{\frac{1}{4}} = \frac{1}{2}$, since $\frac{1}{2} \times \frac{1}{2} = \frac{1}{4}$

  Think How can you use the area model at the right to show $\sqrt{\frac{1}{4}}$?

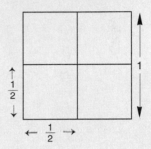

APPLICATION: How Far to the Horizon?

Today's jet passenger planes often fly at an altitude of more than 30 000 feet. On a clear day, how far away is the horizon?

A formula, $d = 1.25\sqrt{a}$, will approximate the distance, d (in miles) for an altitude, a (in feet). At an altitude of 30 000 feet, $d = 1.25\sqrt{30\,000}$. Use a calculator to compute.

$$d = 1.25\sqrt{30\,000}$$
$$\approx 1.25(173.2) \text{ or } 216.5$$

At an altitude of 30 000 feet, the horizon appears about 217 miles away.

ORDER OF OPERATIONS

- The *order of operations* are rules that tell which operations should be done first. Expressions within parentheses must be evaluated first. Then, evaluate all exponents before other operations. Do all multiplications and divisions in order from left to right. Finally, do all additions and subtractions in order from left to right.

EXAMPLE **Evaluate each expression.**

a. $2 + 3 \times 4$

$$2 + 3 \times 4 = 2 + 12$$
$$= 14$$

b. $(2 + 3) \times 4$

$$(2 + 3) \times 4 = 5 \times 4$$
$$= 20$$

c. $4 \times 5 - 14 \div 7$

$$4 \times 5 - 14 \div 7 = 20 - 2$$
$$= 18$$

d. $4(3 + 5) - 2 \times 6$

$$4(3 + 5) - 2 \times 6 = 4(8) - 12$$
$$= 32 - 12$$
$$= 20$$

e. $(2^3 \times 3^2) \div 6 + 8$

$$(2^3 \times 3^2) \div 6 + 8 = (8 \times 9) \div 6 + 8$$
$$= 72 \div 6 + 8$$
$$= 12 + 8$$
$$= 20$$

Think Why do the expressions $3 \times 5 + 4$ and $3(5 + 4)$ have different values, even though they have the same numbers and operations?

APPLICATION: **A Healthy Heart**

The expression $110 + \blacksquare \div 2$ is used to estimate a person's normal blood pressure. In this expression, $\blacksquare$ stands for the person's age in years. Use this expression to estimate the blood pressure of a person who is 36 years old.

Evaluate the expression $110 + 36 \div 2$.

$$110 + 36 \div 2 = 110 + 18$$
$$= 128$$

An estimate of the normal blood pressure is 128.

GRAPHING ORDERED PAIRS

- The *rectangular, or Cartesian, coordinate system* is used to locate points in a plane. The coordinate system is formed by the intersection of two perpendicular number lines that meet at their zero points. The horizontal number line is called the *x-axis,* and the vertical number line is called the *y-axis.* These *axes* separate the plane into four regions called *quadrants.*

- Each point in the plane can be named by an **ordered pair** of numbers. The first number is a measure along the x-axis, while the second number is a measure along the y-axis.

EXAMPLE A **Graph each point on the coordinate plane.**

  a.   A (3, 4)
  b.   B (-1,2)
  c.   C (-2, -4)
  d.   D (5, -5)

EXAMPLE B **In the graph at the right, name the ordered pairs for each point.**

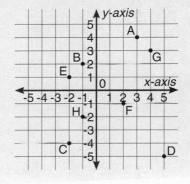

  e.   E   The ordered pair is (-2,1).
  f.   F   The ordered pair is (2, -1).
  g.   G   The ordered pair is (4, 3).
  h.   H   The ordered pair is (-1, -2).

Think   Suppose the first number of an ordered pair is negative. In which quadrants might the point be graphed?

APPLICATION: Measures in Geometry

Transformations are movements of geometric figures. One transformation commonly used is a slide, or translation, of a figure. Draw a square on a coordinate grid and find the coordinates of the vertices. Then draw the translation image 3 units to the left and 4 units down. What are the coordinates of the vertices after the translation?
Before the translation, the coordinates of the vertices are (2,1), (4,1), (2,3), and (4,3).
After the translation, the coordinates of the vertices are (-1, -3), (1, -3), (-1, -1), and (1,-1).

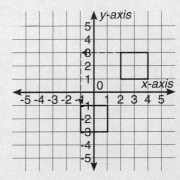

Concepts and Applications

IDEAS ABOUT SURVEYS

- A *survey* can be used to find out how people feel about a certain item or idea. It is not necessary to survey everyone. You can predict responses of an entire *population* by surveying a representative *sample* of that population. A sample is called *random* if the members of the sample are selected purely on the basis of chance. Choosing a random sample may reduce possible *bias,* which happens when the chance of each response is not arbitrary.

EXAMPLE   The members of Student Council wanted to find out students' favorite entertainment: concerts, movies, or dances. One member surveyed a small group of people standing in line at a movie theater. Is this a random sample?

This is not a random sample because those in line at the movie theater may already prefer movies. Also, the people surveyed may not all be students.

> *Think* How might you use a random sample to predict the results of a class election?

- The pieces of information you gather are called the *data.*

APPLICATION: Market Research

Suppose you work for an advertising agency that needs to find out what TV viewers remember from a particular commercial they want to show on TV. What kind of survey data might you collect to answer this question?

You could design a survey to check on what each viewer can remember after seeing the 30-second commercial once. Some questions might be:
a. What product was advertised?
b. What do you remember about the product?
c. Where can you buy the product?
d. Would you buy the product? Why or why not?
e. How could the commercial be changed to be more interesting to you?

CONCEPTS AND APPLICATIONS

DATA TABLES

- From responses to surveys, the data can be presented in data tables. One type of data table used in statistics is a *frequency table.* A frequency table tells how many times each piece of data occurs in a set of information.

EXAMPLE   Make a frequency table for the data below.

Magazine Subscriptions Sold by Swim Club Members

70	74	12	34	23	78	45	32	55	51
89	43	32	11	25	62	43	78	70	72

First, find the least and greatest number of subscriptions. The least is 11 and the greatest is 89.

Then, use intervals to group the data. One choice is to use 10 as the interval.

Finally, make a table with three columns. Write the intervals and tally the numbers of subscriptions.

Subscriptions	Tally	Frequency
10–19	||	2
20–29	||	2
30–39	|||	3
40–49	|||	3
50–59	||	2
60–69	|	1
70–79	⁙|	6
80–89	|	1

APPLICATION: Food Service

The manager of a fast food restaurant had to report the usual time it took for customers to be served after placing an order. She surveyed a random sample of customers and organized the data in the frequency table below.

Serving Time (seconds)	Tally	Frequency
0–60	||	2
61–120	⁙ ⁙	10
121–180	⁙ ⁙ ⁙	15
181–240	⁙ ⁙ ⁙ |||	18
241–300	⁙ ⁙ |	11

Look in the table for the time that has the most tally marks. The usual amount of time it took to serve a customer was between 181 and 240 seconds or between 3 and 4 minutes.

Concepts and Applications

Ratios and Percents

- A *ratio* is a fraction or decimal that compares two numbers. Sometimes a ratio is written a:b and read "a to b". The comparison in a ratio can represent a *rate,* such as miles per hour.

EXAMPLE A   Name the ratio for each situation. Then name another equal ratio.

a.   8 blouses, 5 skirts $\qquad\qquad \frac{8}{5} = \frac{16}{10}$

b.   3 red flowers, 4 yellow flowers $\qquad \frac{3}{4} = \frac{15}{20}$

c.   \$39 for 3 compact discs $\qquad\quad \frac{39}{3} = \frac{13}{1}$

d.   9 deaths per 1 000 people $\qquad \frac{9}{1\,000} = \frac{900}{100\,000}$

- A percent is a special ratio where the second number is 100. Percents are written with the % symbol. For example, $75\% = \frac{75}{100}$.

EXAMPLE B   Express each ratio as a percent.

a. $\frac{33}{100}$ $\qquad\qquad\qquad \frac{33}{100} = 33\%$

b.   62.5 out of 100 $\qquad \frac{62.5}{100} = 62.5\,\%$

c.   15 per 100 $\qquad\qquad \frac{15}{100} = 15\%$

d.   1 out of 2 $\qquad\qquad \frac{1}{2} = \frac{50}{100}$ or 50%

APPLICATION: Sports

The best player on the girls' basketball team made 12 out of 20 free throws so far this season. The best player on the boys' basketball team has made 14 out of 25 free throws so far. Which player has made the greater percent of free throws?

Express each ratio as a percent and compare.

Girls' Team: 12 out of 20 $\qquad$ Boys' Team: 14 out of 25

$\frac{12}{20} = \frac{60}{100}$ or 60% $\qquad\qquad \frac{14}{25} = \frac{56}{100}$ or 56%

Since 60% is greater than 56%, the player on the girls' team has made the greater percent of free throws.

PROPORTIONAL REASONING

- *Proportional reasoning* involves the ideas of a mathematical proportion. *Proportions* use equivalent ratios. Some situations that involve proportions include *comparisons* and *rates*.

EXAMPLE A   At a party, the ratio of girls to boys was 5 to 3. How many girls and boys could possibly be at the party?

This is an example of a comparison. Other ratios that are equal to 5 to 3 are 10 to 6, 15 to 9, and 20 to 12. Each new ratio is equal to the ratio 5:3. For example, 10 girls to 6 boys can be grouped as two sets of 5 girls and 3 boys, as illustrated below.

```
GGGGGGGGGG              (GGGGG) (GGGGG)
  BBBBBB        ——→        BBB      BBB
```

There could be 8, 16, 24, 32, and so on girls and boys at the party.

EXAMPLE B   When shopping for blank audio tapes, Rita found three ways to buy them. Which is the best buy?

Option A: 3 for $8.85
Option B: 5 for $12.45
Option C: 1 for $2.95

This is an example of a rate. Usually the best buy has the least cost per item, assuming items have the same quality. To compare options A, B, and C, you need to find the cost per tape.

Option A

$$\frac{\$8.85}{3} = \frac{?}{1}$$

Since $8.85 \div 3 = 2.95$, the cost per tape is $2.95. Notice this is also the cost when buying a single tape in Option C.

Option B

$$\frac{\$12.45}{5} = \frac{?}{1}$$

Since $12.45 \div 5 = 2.49$, the cost per tape is $2.49. This is less than $2.95. Therefore, the package of 5 is the best buy.

> *Think* Why might Rita choose to buy the package of 3 tapes, even though the package of 5 tapes is the best buy?

- *Population density* refers to the number of individuals per unit of area. A large density means there are many individuals per unit of area. A population density is a rate, so problems involving this idea are often solved using proportional reasoning.

APPLICATION: Home on the Range

Brett is interested in becoming a wildlife manager. He found that surveys are done by wildlife managers to count the number of deer in various habitats. Study the data below. Find the density of the deer populations from the data. Use the density to predict how many deer are in each region.

Region (land area, square miles)	Deer Counted	Survey Area (square miles)
Central Wyoming (25 000)	214	23
Western Minnesota (6 000)	97	40
Northern Alabama (18 000)	223	64
Eastern Colorado (15 000)	114	17
Northern New York (9 000)	55	92
Western Texas (65 000)	85	283
Southern Florida (11 000)	36	30

You can find the population density from the ratio, $\frac{deer\ counted}{survey\ area}$. For central Wyoming, $\frac{214}{23} = 214 \div 23$ or about 9.3 deer per square mile. Therefore, in the Central Wyoming area, the predicted number of deer is about $25\ 000 \times 9.3$ or 232 500 deer. The chart can be extended as shown below.

Region (land area, square miles)	Deer Counted	Survey Area (square miles)	Density	Predicted Number
Central Wyoming (25 000)	214	23	9.3	232 500
Western Minnesota (6 000)	97	40	2.4	14 400
Northern Alabama (18 000)	223	64	3.5	63 000
Eastern Colorado (15 000)	114	17	6.7	100 500
Northern New York (9 000)	55	92	0.6	5 400
Western Texas (65 000)	85	283	0.3	19 500
Southern Florida (11 000)	36	30	1.2	13 200

MEASUREMENT IDEAS

- When you *measure* a quantity, you want to find how many units the quantity has. To measure length, for example, you might use inches, feet, centimeters, or meters, to name a few units. Before you measure, it is important to decide what 1 unit will represent.

EXAMPLE **Use a ruler to find the length, in centimeters and inches, of the segment below.**

With a centimeter ruler, you should find the length is about 10 centimeters. With an inch ruler, you should find the length to be about $3\frac{7}{8}$ inches.

> *Think* Why did you find different measures for the same segment?

- Measures will always have an error. *Errors of measurement* result from inaccuracies in the measuring instrument, from errors in using it, and from mistakes in recording or using the measures. When measuring, it is best to report a *possible error.* Sometimes, scientists use an interval based on the size of the smallest sub-unit on the measuring tool. No matter how it is stated, it is important to remember that any real measure will never be exact.

- *Scale drawings* use proportional reasoning to find the length of very long or very short segments. To reduce or magnify the size while keeping the shape similar to the original, you must apply a specific ratio called the *scale* to the measure of each part.

APPLICATION: Scale Drawings

Suppose that the segment above was drawn to connect two cities on a map. How would you find the actual distance between the two cities?

By looking at the legend of the map, you would find the map scale. Assume the map scale is pictured below.

25 mm = 100 miles

Along the edge of a paper, you could copy the segment for the scale. Then you could see how many times this "fits" into the segment above. Try this. You should find that it fits about 4 times. You can also solve the proportion $\frac{2.5}{10} = \frac{100}{?}$.
Using either method, the estimated length is 400 miles.

FINDING AREAS

- *Area* is a measure of the surface enclosed by a figure. The simplest regions are inside polygons such as squares, rectangles, triangles, or parallelograms, which have segments as sides. Other regions, such as the inside of circles or ellipses, may have curves as boundaries.
- To measure an area, you must first decide on a ***unit area.*** Most often, the unit area is a square with sides of measure 1; for example, 1 centimeter, 1 inch, or 1 kilometer. Such a unit is called 1 ***square unit;*** for example, 1 square centimeter, 1 square inch, or 1 square kilometer. To find the area of a region, you find how many square units would be needed to cover the given region.

EXAMPLE **Find the area of the six-sided region below on the left, using 1 square centimeter as the square unit.**

To find the area of the hexagon, find how many of the unit area squares would be needed to cover the hexagon.

The grid makes it easy to see that the area is 8.5 square centimeters.

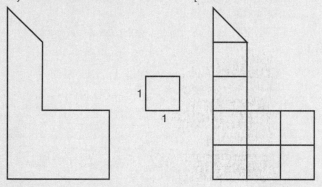

Think For what situations in your life do you need to find the area of a region?

Proportional reasoning can be used to find areas of very large or very small regions.

APPLICATION: **Surface Area of Buildings**

Suppose the figure above is a scale drawing of the side of a large building. Each unit square in the drawing represents an actual square that has sides of 20 meters. What is the area, in square meters, of the side of the building?

Using the scale 20:1, you can reason that the unit square above is equivalent to a square on the building with sides of 20 meters. Each building square contains 20×20 or 400 square meters. From the scale drawing, you know there are 8.5×400 or 3 400 square meters of area on the side of the building.

FINDING VOLUMES

- **Volume** is a measure of the space occupied by a solid. The simplest spatial regions are contained in cubes, pyramids, and rectangular solids, which have polygons as faces and segments as edges. More complicated regions, such as spheres and cylinders, may have curved edges.
- To measure a volume, you must first decide on a **unit volume.** Most often, the unit volume is a cube with edges of measure 1; for example, 1 centimeter, 1 inch, or 1 kilometer. Such a unit is called 1 **cubic unit;** for example, 1 cubic centimeter, 1 cubic inch, or 1 cubic kilometer. To find the volume of a space, you find how many cubic units would be needed to fill the space.

EXAMPLE **The drawing below on the left represents a three-dimensional solid. Find its volume, using 1 cubic centimeter as the cubic unit.**

The spatial region inside the solid could be filled with cubes and half-cubes.

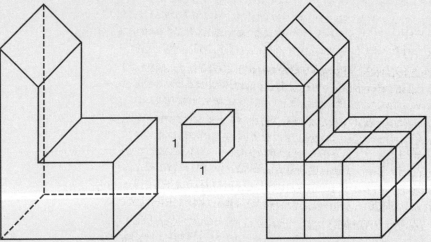

The grid allows you to see that the volume is 25.5 cubic centimeters.

> *Think* For what situations in your life do you need to find the volume of a space?

APPLICATION: Volume of Buildings

Suppose the solid represented above is a model of a very large building. Each unit cube represents an actual cube that has sides of 20 meters. What is the volume, in cubic meters, of the building?

You can use the scale to find that each cube in the building has a volume of $20 \times 20 \times 20$ or 8 000 cubic meters. From the scale drawing, you know the volume of the model is 25.5 cubic units. Therefore, the volume of the building is $25.5 \times 8\ 000$ or 204 000 cubic meters.

Concepts and Applications

EXPONENTS

• *Exponents* are used to abbreviate a repeated multiplication. The number used as the repeated factor is called the *base*, and the exponent tells how many times the base is used as a factor in the multiplication. You can say that the base is raised to a *power* of the exponent.

EXAMPLE A   Write each product using exponents. Then find each product.

a. $2 \times 2 \times 2$ $2 \times 2 \times 2 = 2^3$ or 8

b. $10 \times 10 \times 10 \times 10$ $10 \times 10 \times 10 \times 10 = 10^4$ or 10 000

c. $2 \times 2 \times 2 \times 3 \times 3$ $2 \times 2 \times 2 \times 3 \times 3 = 2^3 \times 3^2 = 8 \times 9$ or 72

Think What is the value of 10^1?

To multiply or divide exponential numbers with like bases, you can use the definition of exponents.

EXAMPLE B   Find each product. Write each product using exponents.

a. $2^2 \times 2^3$

     $2^2 \times 2^3 = (2 \times 2) \times (2 \times 2 \times 2)$

            $= 2^5$

b. $4^3 \times 4^4$

     $4^3 \times 4^4 = (4 \times 4 \times 4) \times (4 \times 4 \times 4 \times 4)$

            $= 4^7$

Notice that in each case you can find the product by adding exponents.

    $2^2 \times 2^3 = 2^{2+3}$ or 2^5 $4^3 \times 4^4 = 4^{3+4}$ or 4^7

EXAMPLE C   Find each quotient. Write each quotient using exponents.

a. $2^5 \div 2^2$

     $2^5 \div 2^2 = \dfrac{2^5}{2^2}$

          $= \dfrac{\cancel{2} \times \cancel{2} \times 2 \times 2 \times 2}{\cancel{2} \times \cancel{2}}$

          $= 2 \times 2 \times 2$ or 2^3

b. $7^4 \div 7^2$

     $7^4 \div 7^2 = \dfrac{7^4}{7^2}$

          $= \dfrac{\cancel{7} \times \cancel{7} \times 7 \times 7}{\cancel{7} \times \cancel{7}}$

          $= 7 \times 7$ or 7^2

Notice that in each case you can find the quotient by subtracting exponents.

    $2^5 \div 2^2 = 2^{5-2}$ or 2^3 $7^4 \div 7^2 = 7^{4-2}$ or 7^2

CONCEPTS AND APPLICATIONS

- *Negative exponents* are often used to represent some numbers between 0 and 1. Consider the quotient $3^2 \div 3^5$. The numbers have the same base, so by subtracting the exponents, the quotient is 3^{2-5} or 3^{-3}. In simplest form, the fraction $\frac{3^2}{3^5}$ is equal to $\frac{1}{3^3}$. Since $3^2 \div 3^5$ cannot have two different values, you can conclude that 3^{-3} is equal to $\frac{1}{3^3}$.

EXAMPLE D Express each negative exponent with a positive exponent. Then find the value without using exponents.

a. 5^{-3} $\qquad$ $5^{-3} = \frac{1}{5^3}$ or $\frac{1}{125}$

b. 10^{-1} $\qquad$ $10^{-1} = \frac{1}{10^1}$ or $\frac{1}{10}$

c. 2^{-4} $\qquad$ $2^{-4} = \frac{1}{2^4}$ or $\frac{1}{16}$

- People who deal regularly with very large or very small numbers use *scientific notation.* When a number is expressed using scientific notation, it is written as a product of a factor and a power of 10, where the exponent is either a positive or negative integer. The factor must be greater than or equal to 1 and less than 10. For example, 2.56×10^3 is written in scientific notation. A numeral such as 2 560 is written in *standard form.*

EXAMPLE E Write the standard form for each number.

a. 4.356×10^2 $\qquad$ $4.356 \times 10^2 = 435.6$
b. 1.7×10^6 $\qquad$ $1.7 \times 10^6 = 1\ 700\ 000$
c. 2.63×10^{-1} $\qquad$ $2.63 \times 10^{-1} = 0.263$
d. 1.005×10^{-3} $\qquad$ $1.005 \times 10^{-3} = 0.01005$

Think What pattern is there between the exponent and the number of places the decimal moves?

APPLICATION: **Astronomy**

The closest approach of a planet to the sun is called its perihelion. In 1991, the planet Pluto was near its perihelion, which is 2 762 000 000 miles. At the same time, Neptune's perihelion was 2.766×10^9 miles. At that time in history, which of these planets was the ninth planet from the sun?

You need to compare Pluto's distance, 2 762 000 000, with Neptune's distance, 2.766×10^9. In standard form, 2.766×10^9 is 2 766 000 000.

Neptune was a greater distance from the sun and therefore was the ninth planet from the sun.

POLYGONS AND POLYHEDRA

- A *polygon* is a 2-dimensional shape in a plane. It is flat and has **sides** that are line segments. The endpoints of the sides are called its **vertices.** Each endpoint is named with a capital leter, and the polygon can be named using every endpoint letter.

EXAMPLE Identify the name and type of polygon shown below.

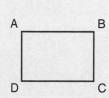 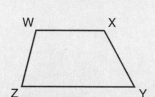 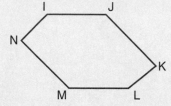

   a. Polygon ABCD is a quadrilateral that appears to be a rectangle.
   b. Polygon WXYZ is a quadrilateral that appears to be a trapezoid.
   c. Polygon IJKLMN is a hexagon.

- A *polyhedron* is a 3-dimensional solid. It has flat surfaces called *faces.* All faces are polygons. Two faces intersect in a line segment called an *edge.* Three or more edges intersect in a point called a *vertex* of the polyhedron. One of the most familiar polyhedra is called a *cube.*

APPLICATION: Crystals

The salt we use to flavor foods comes from the mineral salt. Table salt is in the form of clear crystals shaped like cubes. The diagram at the right represents a salt crystal. Describe several characteristics of a cube.

A cube is a 3-dimensional solid with six faces. Each face is a square. The square faces intersect in edges that are all the same length. There are 12 edges. Adjacent faces are perpendicular, and thus form a right angle.

RIGHT RECTANGULAR PRISMS

• Most of the boxes you find in a grocery store are examples of **right rectangular prisms.** A **prism** is a polyhedron in which at least one pair of faces are parallel and congruent. These are called **bases** and are used to name the prism. Two examples of prisms are shown at the right.

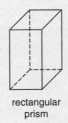 

rectangular prism   triangular prism

Recall that **volume** is the measure of the space occupied by a solid. The prism model at the right was built with 24 cubes. If each cube is a centimeter cube, the volume of the prism is 24 cubic centimeters.

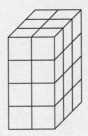

The dimensions of the prism are 2 centimeters by 3 centimeters by 4 centimeters. Notice that 24 is equal to the product $2 \times 3 \times 4$. These and other examples suggest that you can find the volume of a right rectangular prism by multiplying its length, width, and height. Therefore, $V = \ell wh$.

EXAMPLE Find the volume of a right rectangular prism with length 15 inches, width 13 inches, and height 17 inches.

$V = \ell wh$
$V = 15 \times 13 \times 17$ or $3\ 315$
The prism has a volume of 3 315 cubic inches.

Think What is the formula for the volume of a cube?

APPLICATION: Manufacturing

A leading manufacturer of sugar cubes packs a rectangular box so that there are six cubes along one edge, eleven cubes along a second edge, and three cubes along the third edge. How many sugar cubes are in the box? Are there any other arrangements that are possible?
$V = \ell wh$
$V = 6 \times 11 \times 3$ or 198
There are 198 sugar cubes in the box.

It is also possible to arrange 198 cubes using the dimensions $2 \times 3 \times 33$, $3 \times 3 \times 22$, or $2 \times 9 \times 11$.

SURFACE AREA

- The *surface area* is the sum of the areas of all the surfaces or faces of a 3-dimensional solid. The surface area of a right rectangular prism with length ℓ, width w, and height h is $2\ell w + 2wh + 2\ell h$.

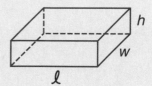

EXAMPLE A   **Find the surface area (S.A.) of a rectangular prism that is 3 inches by 4 inches by 5 inches.**

$S.A. = 2\ell w + 2wh + 2\ell h$
$S.A. = (2 \times 3 \times 4) + (2 \times 4 \times 5) + (2 \times 3 \times 5)$
$\qquad = 24 + 40 + 30$
$\qquad = 94$

The surface area is 94 square inches.

- A *cylinder* is a 3–dimensional solid with a curved surface. Most of the cylinders we see are called **right circular cylinders.** The bases are two parallel, congruent circles.

As with prisms, you find the surface area of the cylinder by finding the area of the two bases and adding the area of the side. You can see that the top and the bottom of a cylinder are in the shape of a circle.

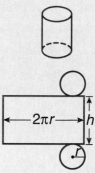

The area of a circle of radius r is πr^2. Since there are two circles, their area is $2\pi r^2$. If you take the curved part of the can and unfold it, you see the curved surface is in the shape of a rectangle. The rectangle has sides of length h and the circumference of the circle, $2\pi r$. Thus, the area of the rectangle is $2\pi rh$. The total surface area is $2\pi r^2 + 2\pi rh$.

EXAMPLE B   **Find the surface area of a 48-ounce juice can. When measured, the diameter of the can is about 10.8 centimeters and the height is about 17.7 centimeters.**

$S.A. = 2\pi r^2 + 2\pi rh$
$S.A. \approx 2 \times 3.14 \times (5.4)^2 + 2 \times 3.14 \times 5.4 \times 17.7$
$\qquad \approx 183.1248 + 600.2424$
$\qquad \approx 783.3672$

The surface area is about 783 square centimeters.

CONCEPTS OF MATHEMATICAL PROBABILITY

- When we think about situations involving chance, we are dealing with the notion of probability. The **probability** of an event occurring is the ratio of the number of favorable outcomes to the number of possible outcomes. With any event, we must first tell what outcomes (or results) are possible. This is called the **sample space.** Then we identify which of the outcomes in the sample space are favorable, or desirable, outcomes. This ratio, based on mathematical principles, is called the **theoretical probability.** We often write P(favorable outcome) to denote the probability of a specific outcome.

EXAMPLE A   **Many games use a numbered cube, called a die. It is assumed that when you roll a die, any one of the six faces has an equal chance of being the top face. What is the probability that you get an even number when you roll a standard die?**

The favorable outcomes occur when either a 2, 4, or 6 (even number) shows. There are 6 possible outcomes: 1, 2, 3, 4, 5, or 6. The ratio $\frac{3}{6}$ is the number of favorable outcomes to the number of possible outcomes. Thus, the probability of an even number, P(even), is $\frac{3}{6}$ or $\frac{1}{2}$.

- **Experimental probability** is a ratio found by repeating a certain number of trials. The ratio of the number of observed successful outcomes to the number of tries is the experimental probability value. The experimental probability of a certain outcome will not always equal the theoretical probability of the outcome.

EXAMPLE B   **An experiment was conducted where a die was rolled 24 times. The results are shown in the chart at the right. What was the experimental probability of rolling a 1? a 5?**

From the chart we can find the ratio of the number of successful outcomes for a number on the die to the total number of outcomes. This ratio is the experimental probability.

Outcome	Number of Successful Outcomes
1	5
2	3
3	6
4	4
5	2
6	4
Total	24

experimental probability of rolling a 1: $\frac{5}{24}$

experimental probability of rolling a 5: $\frac{2}{24}$ or $\frac{1}{12}$

> **Think** What is the *theoretical probability* of rolling a 1? a 5? How do these probabilities compare to the experimental probabilities from this experiment?

Concepts and Applications

THE FUNDAMENTAL COUNTING PRINCIPLE

The *Fundamental Counting Principle* can often be used to find the number of all possible outcomes in a situation. This principle states that if there are *m* ways of selecting an item from set A and *n* ways of selecting an item from set B, then there are *m* × *n* ways of selecting an item from set A and an item from set B.

EXAMPLE Find the total number of possible outcomes for the spinners below.

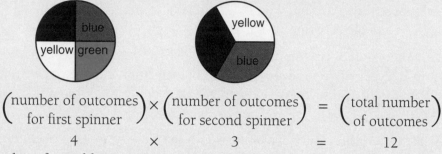

$$\left(\begin{array}{c}\text{number of outcomes}\\\text{for first spinner}\end{array}\right) \times \left(\begin{array}{c}\text{number of outcomes}\\\text{for second spinner}\end{array}\right) = \left(\begin{array}{c}\text{total number}\\\text{of outcomes}\end{array}\right)$$

$$4 \qquad \times \qquad 3 \qquad = \qquad 12$$

The total number of possible outcomes is 12.

APPLICATION: Radio and Television Stations

Radio and television stations in the United States have names that begin with either a K or a W. Some stations have a 3-letter name, but most stations licensed after 1927 have used 4 letters. How many 3-letter names are possible? 4-letter names?

There are 2 choices for the first letter, K or W. There are 26 choices for each of the second and third letters. Use the Fundamental Counting Principle.

$2 \times 26 \times 26 = 1\ 352$

There are 1 352 possible 3-letter names.

To find the number of 4-letter names, use the Fundamental Counting Principle again.

$2 \times 26 \times 26 \times 26 = 35\ 152$

There are 35 152 possible 4-letter names.

Think How are the identification codes on the car license plates in your state made? Usually it is a number, or a combination of letters and numbers. Use the Fundamental Counting Principle to find how many possible car license plates can be used in your state.

Multiplying Probabilities

• You can find the probability of two independent events by *multiplying the probabilities* of each event.

EXAMPLE   One bag contains 3 red and 4 white balls. A second bag contains 6 yellow and 3 green balls. One ball is drawn from each bag. Find the probability of drawing a red ball from the first bag and a yellow ball from the second bag.

There are a total of 7 balls in the first bag.
$P(\text{red ball}) = \frac{3}{7}$

There are a total of 9 balls in the second bag.

$P(\text{yellow ball}) \quad = \frac{6}{9} \text{ or } \frac{2}{3}$

$P(\text{red and yellow}) = \frac{3}{7} \times \frac{2}{3}$

$\qquad\qquad\qquad\quad = \frac{2}{7}$

The probability that a red ball will be drawn from the first bag and a yellow ball from the second bag is $\frac{2}{7}$.

APPLICATION: Clothing

In one drawer, Tei has 2 pairs of brown socks, 3 pairs of black socks, and 4 pairs of blue socks. In another drawer, he has 3 red sweaters, 2 brown sweaters, and 2 blue sweaters. Suppose Tei makes a selection from each drawer without looking. What is the probability that he will have brown socks and a brown sweater?

$P(\text{brown socks}) \quad = \frac{2}{9}$

$P(\text{brown sweater}) = \frac{2}{7}$

$P(\text{brown socks and brown sweater}) = \frac{2}{9} \times \frac{2}{7}$

$\qquad\qquad\qquad\qquad\qquad\qquad\qquad = \frac{4}{63}$

The probability that he will have brown socks and a brown sweater is $\frac{4}{63}$.

Concepts and Applications

Mathematical Odds

- The *odds in favor* of an event is the ratio of the total number of favorable outcomes to the total number of unfavorable outcomes. The *odds against* an event is the reciprocal of the odds in favor.

EXAMPLE   **A die is rolled. Find the odds in favor of rolling a number greater than 4. Then, find the odds against rolling a number greater than 4.**

The numbers on a die greater than 4 are 5 and 6.
The numbers on a die not greater than 4 are 1, 2, 3, and 4.
So, there are 2 favorable outcomes and 4 unfavorable outcomes.

$$\text{odds in favor} = \frac{2}{4} \text{ or } \frac{1}{2}$$

$$\text{odds against} = \frac{4}{2} \text{ or } \frac{2}{1}$$

The odds of rolling a number greater than 4 are $\frac{1}{2}$, and the odds against are $\frac{2}{1}$.

APPLICATION: Entertainment

The eighth-grade class at Clinton Middle School is planning an end-of-year dance. The dance committee decides to purchase 20 door prizes to be awarded randomly at the dance. If 180 students are expected to attend the dance, what are the odds of winning a door prize? Explain.

$$\text{number of favorable outcomes} = 20$$

$$\text{number of unfavorable outcomes} = 180 - 20 = 160$$

$$\text{odds in favor} = \frac{20}{160} \text{ or } \frac{1}{8}$$

The odds in favor of winning a door prize are $\frac{1}{8}$. This means that for every student that wins a prize, 8 students do not.

USING A PROTRACTOR

- A *protractor* is a tool for measuring angles. The scale on the protractor corresponds to the degree measure of an angle. Each unit is called a *degree.* There are 360° in a circle, 180° in a half circle, and 90° in a quarter circle.

EXAMPLE A   Use a protractor to measure the angle at the right.

Place the protractor on the angle so that the center is on the vertex of the angle and the 0° line lies on one side of the angle. Then follow the scale from the 0° point to the point where the other side of the angle meets the scale. The angle's measure is 50°.

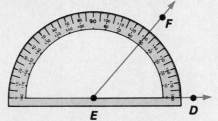

- You can use a protractor to *make a circle graph.* A circle graph is used to compare parts of a whole.

EXAMPLE B   Jason recorded the number of hours he spent on homework in one week. Use this data to make a circle graph.

Homework (in hours)	
Math	4
English	3
History	3
Science	2
Total	12

Find the ratio that compares the time spent on each subject to the total time.

Math: $\frac{4}{12}$ or $\frac{1}{3}$ History: $\frac{3}{12}$ or $\frac{1}{4}$

English: $\frac{3}{12}$ or $\frac{1}{4}$ Science: $\frac{2}{12}$ or $\frac{1}{6}$

Find the number of degrees for each section of the graph by writing a proportion. Remember, there are 360° in a circle.

Math: $\frac{1}{3} = \frac{?}{360°}$ The angle is 120°.

English: $\frac{1}{4} = \frac{?}{360°}$ The angle is 90°.

History: $\frac{1}{4} = \frac{?}{360°}$ The angle is 90°.

Science: $\frac{1}{6} = \frac{?}{360°}$ The angle is 60°.

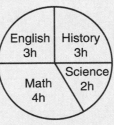

Then use a protractor to draw each angle in a circle graph.

Concepts and Applications

ELEMENTARY GEOMETRIC CONCEPTS

- A **line** is a never-ending straight path. A **line segment** is a straight path between two endpoints. A **ray** is a part of a line that extends indefinitely in one direction. An **angle** is two rays with a common endpoint.
- **Parallel lines** are lines in the same plane that do not intersect. Segments and rays can also be parallel. **Perpendicular lines** are two lines in the same plane that intersect to form right angles (90° angles). Segments and rays can also be perpendicular.
- Points are named with capital letters. Letters are used to name parts in the geometric figure in the example below.

EXAMPLE A Use the figure at the right to name the parts listed below.

    a. two lines
        Possible lines are $\overleftrightarrow{AC}$, $\overleftrightarrow{DE}$, and $\overleftrightarrow{EF}$.

    b. three line segments
        Possible line segments are $\overline{AB}$, $\overline{BC}$, $\overline{AC}$, $\overline{BE}$, $\overline{BD}$, $\overline{DE}$, $\overline{BF}$, and $\overline{EF}$.

    c. two rays
        Possible rays are $\overrightarrow{BD}$, $\overrightarrow{BC}$, $\overrightarrow{BA}$, $\overrightarrow{BE}$, $\overrightarrow{ED}$, $\overrightarrow{FE}$, $\overrightarrow{EF}$, $\overrightarrow{DE}$, $\overrightarrow{AC}$, and $\overrightarrow{CA}$.

    d. two parallel lines
        Lines AC and EF appear to be parallel.

    e. two perpendicular line segments
        Segments $\overline{BF}$ and $\overline{AB}$, $\overline{BF}$ and $\overline{BC}$, $\overline{AC}$ and $\overline{BF}$, and $\overline{BF}$ and $\overline{EF}$ appear to be pairs of perpendicular segments.

    f. three angles
        Possible angles are $\angle\,ABD$, $\angle\,CBD$, $\angle\,ABC$, $\angle\,ABE$, $\angle\,ABF$, $\angle\,CBE$, $\angle\,CBF$, $\angle\,EBD$, $\angle\,EBF$, and $\angle\,DBF$.

- The **bisector of a line segment** is a line that divides the segment into two congruent segments. (Congruent means having equal measure.) The **midpoint of a segment** is the point at which the bisector intersects the segment. The **perpendicular bisector** is a bisector that is also perpendicular to the segment.
- The **bisector of an angle** is a ray through the interior of an angle that divides the angle into two congruent angles.

EXAMPLE B Name the bisector of the angle at the right.

The bisector of the angle is $\overrightarrow{AD}$ because it is a ray in the interior of $\angle\,BAC$ that divides $\angle\,BAC$ into two congruent angles.

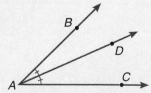

CONCEPTS AND APPLICATIONS

TYPES OF POLYGONS

- A *polygon* is a simple closed figure in a plane formed by three or more line segments. A **vertex** is a point where two sides of the polygon intersect. A polygon is named using the letters that label its vertices.

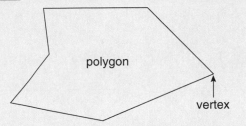

Think How are the number of vertices and the number of sides of a polygon related?

- A *triangle* is a polygon that has 3 sides. A **quadrilateral** is a polygon having 4 sides. A **pentagon** and a **hexagon** have 5 and 6 sides, respectively. An **n-gon** has n sides.

- There are many types of triangles. Some are based on the size of its angles. An *acute triangle* has three angles that each measure less than 90°. A **right triangle** has a 90° angle. An **obtuse triangle** has an angle that measures greater than 90°.

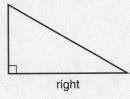
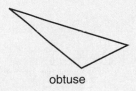

acute   right   obtuse

Think Why do all triangles have at least two acute angles?

- Some types of triangles depend on the length of its sides. A *scalene triangle* has no congruent sides. An *isosceles triangle* has at least two congruent sides. An *equilateral triangle* has three conguent sides.

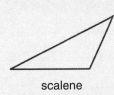

scalene   isosceles   equilteral

EXAMPLE A   Name the types of triangles in the figure at the right.

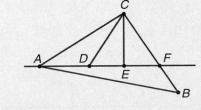

right triangles: *ACB, ACF, CED, CEF, CEA*
obtuse triangles: *ADC, AFB*
acute triangles: *CDF*
scalene triangles: *ABF, ACD*
isosceles triangles: *ACB*
equilateral triangles: *CDF*

Think Can a right triangle also be an equilateral triangle? Why or why not?

Concepts and Applications

- There are many different types of quadrilaterals. A *trapezoid* has exactly one pair of parallel sides. A *parallelogram* has two pairs of parallel sides. A *rectangle* is a parallelogram with four right angles. A *rhombus* is a parallelogram with all sides congruent. A *square* is a parallelogram with four right angles and all sides congruent.

EXAMPLE B   **Name the types of quadrilaterals in the figure below.**

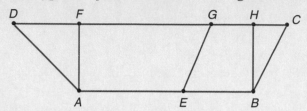

trapezoids: *ABCD, AEGD, ABHD, ABCF, AEGF, EBHG*
parallelograms: *ABHF, EBCG*
rectangles: *ABHF*
rhombuses: *EBCG*
squares: none

> **Think** Is every square a rectangle? Is every rectangle a square?

- The *interior* of a polygon consists of all points inside the polygon. The *exterior* consists of all points outside the polygon. A *diagonal* of a polygon is a line segment that joins two nonconsecutive vertices of a polygon.

> **Think** Must the diagonals of a polygon lie completely in the interior of the polygon? Draw examples to show the possible cases.

APPLICATION: Traffic Safety

There are certain types of traffic signs that always have the same shape. Identify the shape of each sign below and tell what the sign means.

a.   b.   c.   d.

octagon, stop   square, curve   triangle, yield   rectangle, speed limit 50

FORMULAS FOR AREAS

- The **area of a rectangle** with a width w units and length ℓ units is ℓw square units. The **area of a triangle** with base b units and height (altitude) h units is $\frac{1}{2}bh$ square units. The **area of a parallelogram** with base b units and height h units is bh square units. The **area of a trapezoid** with parallel bases a and b units and height h units is $\frac{1}{2}h(a+b)$ square units.

EXAMPLE A A football field is a rectangle 360 feet long and 160 feet wide. Find the area of a football field in square feet.

$A = \ell w$
$\quad = (360)(160)$
$\quad = 57\ 600$
The area of a football field is 57 600 square feet.

EXAMPLE B Find the area of the triangle at the right.

$A = \frac{1}{2}bh$
$\quad = \frac{1}{2}(18)(7)$
$\quad = 63$
The area of the triangle is 63 square meters.

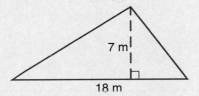

EXAMPLE C What is the area of a parallelogram with sides 12 inches and 8 inches and height 6 inches?

First, draw a figure.

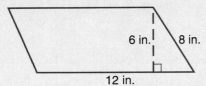

$A = bh$
$\quad = (12)(6)$
$\quad = 72$
The area of the parallelogram is 72 square inches.

EXAMPLE D Find the area of the trapezoid at the right.

$A = \frac{1}{2}h(a+b)$
$\quad = \frac{1}{2}(14)(21 + 17)$
$\quad = 266$
The area of the trapezoid is 266 square centimeters.

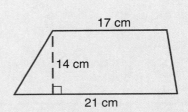

323   *Concepts and Applications*

APPLICATION: Archeology

According to the Guiness Book of Records, the pyramid with the largest base area is the Quetzsalcóatl near Mexico City. Its base is a square and its four faces are isosceles triangles. The base of each triangle is about 604 meters, and the height of each triangle is about 307 meters. Find the area of a triangular face. Then find the surface area.

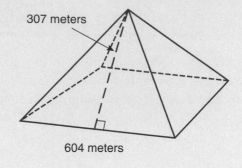

307 meters

604 meters

area of a triangular face $= \frac{1}{2} bh$

$$= \frac{1}{2}(604)(307)$$

$$= 92\,714$$

The area of a triangular face is 92 714 square meters.

area of base $= \ell w$

$$= (604)(604)$$

$$= 364\,816$$

surface area $= 4$(area of a trianglular face) $+$ (area of base)

$$= 4(92\,714) + 364\,816$$

$$= 370\,856 + 364\,816$$

$$= 735\,672$$

The surface area of the pyramid is 735 672 square meters.

APPLICATION: Real Estate

The Wilsons are buying a lot on which they plan to build a house. The shape of the lot is a trapezoid as shown at the right. How many square feet are contained in the lot?

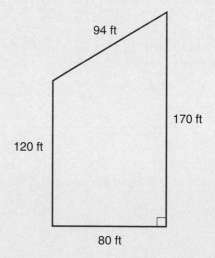

94 ft

170 ft

120 ft

80 ft

$A = \frac{1}{2}h(a + b)$

$$= \frac{1}{2}(80)(120 + 170)$$

$$= \frac{1}{2}(80)(290)$$

$$= (40)(290)$$

$$= 11\,600$$

The area of the lot is 11 600 square feet.

CONCEPTS AND APPLICATIONS

PERIMETER FORMULAS

- The *perimeter of a triangle* is the sum of the measures of the three sides. The *perimeter of a rectangle* is the sum of the measures of the four sides. The *perimeter of an n-gon* is the sum of the measures of the *n* sides.

EXAMPLE A   Find the perimeter of the triangle at the right.

$P = 17 + 14 + 21$
$\quad = 52$

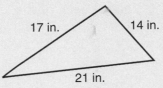

The perimeter of the triangle is 52 inches.

EXAMPLE B   Find the perimeter of a rectangle with a length of 9 feet and a width of 3 feet.

$P = 9 + 3 + 9 + 3$
$\quad = 24$

The perimeter of the rectangle is 24 feet.

APPLICATION: Baseball

Home plate is made of white rubber set into the ground so that it is level with the ground. What is the perimeter of home plate?

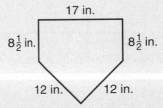

$P = 17 + 8\frac{1}{2} + 8\frac{1}{2} + 12 + 12$
$\quad = 58$

The perimeter of home plate is 58 inches.

CIRCLES

- A *circle* is the set of all points in a plane that are the same distance from a given point called the *center*. A *chord* is a segment with its endpoints on the circle. A *diameter* of a circle is a chord through its center. A *radius* is a segment whose endpoints are the center of the circle and any point on the circle.

EXAMPLE A   Name each of the following in the figure at right.

    a. center: C
    b. chord: $\overline{AB}, \overline{AE}, \overline{AD}, \overline{BE}$
    c. diameter: $\overline{AD}, \overline{BE}$
    d. radius: $\overline{CA}, \overline{CD}, \overline{CE}, \overline{CB}$

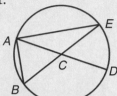

- The *measure of a radius* of a circle is one-half the measure of the diameter $(r = \frac{1}{2}d)$. The *circumference of a circle* is equal to π times its diameter $(C = \pi d)$ or π times twice the radius $(C = 2\pi r)$. The *area of a circle* is equal to π times the square of the radius $(A = \pi r)$.

EXAMPLE B   Find the circumference of the circle at the right.

    $C = 2\pi r$
      $\approx 2(3.14)(9)$ $\pi \approx 3.14$
      ≈ 56.52

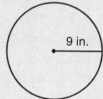

9 in.

The circumference of the circle is about 57 inches.

EXAMPLE C   Find the area of a circle if its radius is 11 meters.

    $A = \pi r^2$
      $\approx (3.14)(11)^2$
      $\approx (3.14)(121)$
      ≈ 379.94 The area of the circle is about 380 square meters.

APPLICATION: Sports

The Houston Astrodome is built in the shape of a circle. Find the circumference and area of the Astrodome's circular floor if its diameter is **214 yards.**

    $C = \pi d$
      $\approx (3.14)(214)$
      ≈ 671.96 The circumference is about 672 yards.

    $r = \frac{1}{2}d$ $A = \pi r^2$
      $= \frac{1}{2}(214)$ $\approx (3.14)(107)^2$
      $= 107$ $\approx (3.14)(11\ 449)$
                        $\approx 35\ 949.86$ The area is about 35 950 square yards.

CONCEPTS AND APPLICATIONS

INTEGERS

- An *integer* is any number from the set {. . . , -4, -3, -2, -1, 0, +1, +2, +3, +4, ...}.

- To *add integers* with the same sign, add their absolute values. Give the result the same sign as the integers. To add integers with different signs, subtract their absolute values. Give the result the same sign as the integer with the greater absolute value.

EXAMPLE A   Solve $a = -1 + (-2)$.

Using counters: Put 1 negative counter on the mat. Add 2 negative counters.

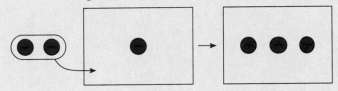

Using arithmetic: $a = -1 + (-2)$ The signs are the same.
     $= -3$ The sum is negative.

EXAMPLE B   Solve $b = -3 + 5$.

Using counters: Put 3 negative counters on the mat. Add 5 positive counters. Remove 3 zero pairs.

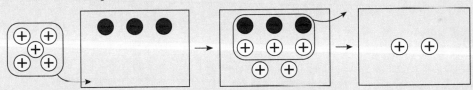

Using arithmetic: $b = -3 + 5$ The signs are different.
      $= 2$ Since $|-3| < |5|$, the sum is positive.

- To *subtract an integer*, add its additive inverse.

EXAMPLE C   Solve $x = -6 - (-3)$.

Using counters: Put 6 negative counters on the mat. Remove 3 of them.

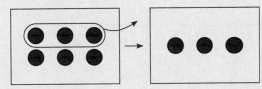

Using arithmetic: $x = -6 - (-3)$
       $= -6 + 3$ To subtract -3, add 3.
       $= -3$

327   *Concepts and Applications*

EXAMPLE D   Solve $y = -6 - 3$ using counters.

Using counters: Put 6 negative counters on the mat. You need to remove 3 positive counters. Since there are none, add three zero pairs. Now you can remove 3 positive counters.

Using arithmetic: $y = -6 - 3$
$$= -6 + (-3) \qquad \text{To subtract 3, add -3.}$$
$$= -9$$

- The **product of two integers** with different signs is negative. The product of two integers with the same sign is positive.

EXAMPLE E   Solve $h = 3(-7)$.

$h = 3(-7)$ The signs are different.
$\;\;= -21$ The product is negative.

EXAMPLE F   Solve $g = -5(-2)$.

$g = -5(-2)$ The signs are the same.
$\;\;= 10$ The product is positive.

- The **quotient of two integers** with the same sign is positive. The quotient of two integers with different signs is negative.

EXAMPLE G   Solve $c = -16 \div (-2)$.

$c = -16 \div (-2)$ The signs are the same.
$\;\;= 8$ The quotient is positive.

EXAMPLE H   Solve $d = -27 \div 3$.

$d = -27 \div 3$ The signs are different.
$\;\;= -9$ The quotient is negative.

Variables and Expressions

- In algebra, a placeholder is called a *variable* because the value can change or vary. We often denote a variable with a letter, such as x.

EXAMPLE A **Explain how each of the following could be a variable.**

a. your height — Through the years, your height has increased, including periods when you had a "growth spurt."

b. hair length — The length of a person's hair increases as it grows, but it decreases when it is cut.

c. pulse — A person's pulse varies according to the rate at which the heart pumps, increasing with exercise and decreasing when quiet.

d. earnings — The earnings of a company or an individual can change from one year to the next.

- In mathematics, we are often interested in how two or more variables interact. The *independent variable* is the variable that you change or vary. The *dependent variable* changes in relation to the independent variable. In an ordered pair, the first value is the independent variable, and the second value is the dependent variable. Thus, an ordered pair has the form (independent, dependent).

EXAMPLE B **For each of these situations, identify ordered pairs to show which are the possible independent and dependent variables.**

a. cost of buying meat
(number of pounds, cost of meat)

b. running to burn calories
(number of miles, number of calories burned)

c. depth of a lake from rainfall
(inches of rain, number of feet of depth)

- An *algebraic expression* is a phrase involving one or more variables, constants, and arithmetic operations. When values are substituted for the variables, the expression becomes an *arithmetic expression,* which can be simplified.

Concepts and Applications

EXAMPLE C   Write an algebraic expression for each situation. Evaluate each expression for some reasonable value.

a. total cost, when each item costs $3 and there is a $2 handling cost

An expression is $3n + 2$, where n represents the number of items. When $n = 22$ items, the total cost is $3(22) + 2$ or $68.

b. points scored by a basketball player from 3-point goals, 2-point goals, and free throws

An expression is $3x + 2y + z$, where x represents number of 3-point goals, y represents number of 2-point goals, and z represents number of free throws. For 3 3-point goals, 8 2-point goals, and 5 free throws, the total number of points is $3(3) + 2(8) + 5$ or 30 points.

c. total number of gallons of gasoline used by a car getting 30 miles per gallon

An expression is $m \div 30$, where m represents the number of miles driven. To go 480 miles, the car would use $480 \div 30$ or 16 gallons of gasoline.

- The **Distributive Property** states that the product $a(b + c)$ is equal to the sum of two products, ab and ac. This idea is useful when simplifying algebraic expressions and solving equations.

EXAMPLE D   **Suppose a rectangle has sides with measures 7 and 5 + 3. Find the area in two different ways.**

Method 1
Using the order of operations,
$7(5 + 3) = 7(8)$ or 56.

Method 2
Find the area of each of the smaller rectangles.
Then add.
$(7 \times 5) + (7 \times 3) = 35 + 21$
$ = 56$

Using either method, the area of the rectangle is 56 square units. This suggests that $7(5 + 3) = (7 \times 5) + (7 \times 3)$.

APPLICATION: At the Movies

In a theater there are two sections with 32 rows each. In the left section, there are 8 seats in each row. In the right section, there are 12 seats in each row. How many seats are in the theater?

The total number of seats can be found in two ways.
$32(8 + 12) = 32 \times 20$ or 640 $32(8 + 12) = (32 \times 8) + (32 \times 12)$
$\phantom{32(8 + 12) = 32 \times 20 \text{ or } 640 \quad 32(8 + 12) } = 256 + 384$ or 640

Using either method, there are 640 seats in the theater.

MATHEMATICAL FUNCTIONS

- A *function* is a set of ordered pairs, (x,y), such that for each value of x there is exactly one value for y.

EXAMPLE A   Determine whether each of the following are functions.

    a. $\{(1,0), (1,1), (1,2)\}$ No; when $x = 1$, y has more than one value.
    b. $\{(0, -1), (1, -2), (2, -3), (3, -4)\}$ Yes
    c. $\{(0.5,0.5), (1,1), (1.5, 1.5)\}$ Yes
    d. $\{-2, -3, -4, -5, -6\}$ No; it is not a set of ordered pairs.
    e. $\{(0,0)\}$ Yes

Sometimes it may be difficult to determine whether there is more than one value of y for each value of x. Often it is simpler to look at a graph. Consider the equation $y = 2x + 1$.

First make a table of values.

x	0	1	2	3
y	1	3	5	7

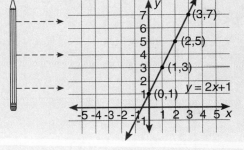

Then graph the ordered pairs and draw a line connecting the points.
Now place your pencil at the left of the graph to represent a vertical line. Slowly move the pencil to the right across the graph.
For each value of x, this vertical line passes through no more than one point on the graph. This is called the **vertical line test.**

EXAMPLE B   Use the vertical line test to determine whether each graph represents a function.

  a. 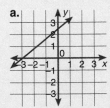   **b.** 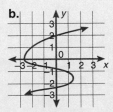   **c.** 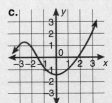   **d.**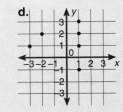

The vertical line test shows that **a** and **c** are functions since, for each value of x, a vertical line passes through no more than one point on the graph. However, **b** and **d** are not functions. In **b**, a vertical line near the y-axis passes through three points. In **d**, a vertical line at $x = 1$ passes through four points.

LINEAR FUNCTIONS AND SLOPE

- A **linear function** is a function whose graph is a nonvertical line. A linear function may be graphed by making a table of values, graphing the ordered pairs, and drawing a line to connect the points.

EXAMPLE A   **Graph each linear function.**

a. $y = x$ **b.** $y = 2x$ **c.** $y = 5x$

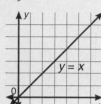 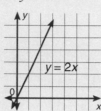 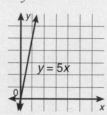

Each of the graphs shown above pass through the origin. However, they vary in steepness. In mathematics, the steepness of a line is called its **slope.**

EXAMPLE B   **Which has a greater slope, $y = 3x$ or $y = \frac{1}{3} x$?**

Graph each function.

 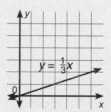

It appears from the graphs that $y = 3x$ has a steeper slope.
So, $y = 3x$ has a greater slope.

> **Think** Can you find a relationship between the steepness of a line and the coefficient of x? Which line has a steeper slope, $y = 2x$ or $y = 6x$? Test your conjecture with a graph.

Solving Equations

- To *solve an equation* means to find those values of the variable or variables that make the equation true. Sometimes, you can use the guess-and-check strategy to solve equations.

EXAMPLE A **Solve the equation $3x + 1 = x - 3$.**

Try different values for x and evaluate each side of the equation. A table may help organize the information.

x	-3	-2	-1	0	1	2	3	4
$3x + 1$	-8	-5	-2	1	4	7	10	13
$x - 3$	-6	-5	-4	-3	-2	-1	0	1

We see that when $x = -2$, both sides of the equation equal -5. Therefore, -2 is a *solution* of the equation.

- In algebra, you will learn to solve equations using inverse properties and properties of equality to write *equivalent equations.* Equivalent equations have the same solution. Some properties are listed below.

Additive Inverse Property	Adding a number and its opposite results in 0.
Multiplicative Inverse Property	Multiplying a number and its reciprocal results in 1.
Addition Property of Equality	Adding the same number to each side of an equation results in an equivalent equation.
Multiplication Property of Equality	Multiplying each side of an equation by the same number results in an equivalent equation.

EXAMPLE B **Solve $3x - 1 = 8$ using properties of inverses and equality.**

$$3x - 1 = 8$$
$$3x - 1 + 1 = 8 + 1 \qquad \text{Add 1 to each side.}$$
$$3x = 9 \qquad \text{Simplify.}$$

$$\tfrac{1}{3}(3x) = \tfrac{1}{3}(9) \qquad \text{Multiply each side by } \tfrac{1}{3}.$$
$$x = 3 \qquad \text{Simplify.}$$
$$\text{The solution is 3.}$$

Think What would be the first step you would use to solve the equation $3x + 2 = 10$?

INTEREST ON MONEY

- To compute the **simple interest** earned, use $I = prt$, where I is the interest, p is the principal amount, r is the rate, and t is the time.

EXAMPLE A A savings account of $100 earns simple interest at 4% annually. Find the amount in the savings account at the end of 5 years.

$$I = prt \qquad\qquad p = 100, \ r = 0.04, \ t = 5$$
$$= 100 \times 0.04 \times 5$$
$$= 20$$

After 5 years, the interest earned is $20. The total amount in the savings account is $100 + $20 or $120.

- To compute **compound interest,** we must increase the principal by the interest earned in each period. This new principal value is used to compute the interest in the next period.

EXAMPLE B Find the amount in the savings account above after 5 years if the interest is compounded each year.

The amount at the end of each year is equal to the amount at the beginning of the year plus 0.04 of the amount at the beginning of the year.
amount at end of year 1: 100 + 0.04(100) = 104
amount at end of year 2: 104 + 0.04(104) = 108.16
amount at end of year 3: 108.16 + 0.04(108.16) = 112.4864
amount at end of year 4: 112.4864 + 0.04(112.4864) = 116.985856
amount at end of year 5: 116.985856 + 0.04(116.985856) = 121.6652902

With interest compounded annually, the amount in the savings account after 5 years is $121.67.

- The amount of money in the account earned by compound interest can be computed directly using the formula $a_n = a_0(1 + r)^n$, where a_n is the accumulated amount in the account after n years, a_0 is the initial deposit, and r is the annual simple interest rate.

EXAMPLE C Use the formula for compound interest to find the amount in the savings account given in Example A.

$a_5 = 100(1 + 0.04)^5$
$\quad = 100(1.04)^5 \qquad$ Use the y^x key on your calculator.
$\quad = 121.6652902$
The amount in the savings account is $121.67.

> **Think** How can the compound interest formula be found from the sequence of steps in Example B? Explain.

CONCEPTS AND APPLICATIONS

GROWTH CURVES

- An **S-curve** is a graph of the typical growth of a population. The graph shows how the size of the population changes over time. It is the graph of the ordered pairs (time, number of individuals) connected by a smooth curve.

EXAMPLE Study the graph of the growth of a population of deer studied in an area designated a "wildlife preserve" in 1948. Why is it called an S-curve?

It is called an S-curve because the shape of the graph looks somewhat like the letter S.

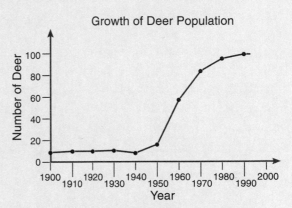

Growth of Deer Population

- There are three distinct phases in the shape of an S-curve. In the early life of a population, called the **lag phase,** we typically see slow growth. The graph is quite flat and "climbs" slowly. The increase in the number of individuals does not change much over time. In the **exponential growth phase,** which follows the lag phase, the population increases rapidly. During this segment of time, the graph "climbs" rapidly as the population grows relatively quickly. The increase in the number of individuals changes quickly over time. In the **stationary phase,** the last of the three stages, the population growth slows down. The size of the population may remain large, but it is constant.

APPLICATION: Wildlife

Study the graph of the growth of a population of deer in the example above.

a. In the graph above, during what time interval was the lag phase? What might explain the population growth of deer during this phase?

From this graph, the lag phase occurred from 1900 to about 1950. During this period, the number of deer remained about the same. Without more specific information about the environment, we must assume that various factors operated to control the number of births and deaths in such a way that the population remained fairly constant in size.

CONCEPTS AND APPLICATIONS

b. During what time interval did the exponential growth phase occur? What might explain the growth of the deer population during this phase?

Rapid growth seemed to begin after 1950 and continue into the 1970s. Since the area was designated a wildlife preserve in 1948, hunting would have been prohibited. Thus, the population reproduced unchecked, resulting in rapid increases in the number of deer.

c. During what time interval did the stationary phase occur? What are some factors which may explain why the deer population growth slowed and stabilized?

It appears that in the 1970s, the growth began to slow, shown by the curve beginning to level off. Thereafter, it appears that the population stabilized because the curve becomes relatively flat. This "zero growth" may have occurred as the environmental factors worked to prevent further increases, yet it supported those deer that were alive.

GLOSSARY/INDEX

Acute triangle, 171-172, 177

Adjacent, 9, 174

Algebra
alternate solutions, 12
axes, 22, 60, 61, 189, 209, 210, 218, 220-221, 235, 245
base, 98
Cartesian coordinate system, 22
coefficient, 230, 237
constant, 230, 237
constant function, 19
equations, 206, 209, 217-244
exponents, 98
expressions, 206
grid, 9-11, 60, 61, 122, 182, 185
horizontal axis, 22-24, 62, 189
intervals, 57, 60 *illus.*, 62
number line, 4, 8, 20, **22**, 29, 138-139
order of operations, 16
ordered pair, 22, 25, 27, 151, 221-244
ordering, 4, 146, 147
origin, 22
prime factorization, **17**
range, 5, 41, 46, 60, 62, 140, 144, 169
reciprocal, **19**, 30
rectangular coordinate system, **22**
scale, **60**, 221, 225
scientific notation, 97-101
sets, 17
variables, **60**, 206
Venn diagrams, **12**, 30
vertical axis, 22, 62, 189

Angle bisector, 171 a ray or a line that divides an angle in half

Angles, 170-171
interior, 227
measure of, 172

Area, 92, **93**, 99, 110, 166, 172, 174-187, 190-192, 202 the number of square units needed to cover a surface
of circles, 188, 190 *illus.*, 190-192, 195
of containers, 168-169
of parallelograms, 175-178
of polygons, 179-180, 183
of rectangles, 174, 176, 186, 195, 203
of squares, 182-183
of the base, 193
of trapezoids, 178-179
of triangles, 175-178, 219
surface area of cylinders, 195-196

Attributes, 111 characteristics that help describe an object

Average, 49-52, 61, 102-103, 137, 169, 180, 189

Axes, 22, 60, 61, 189, 209, 210, 218, 220-221, 235, 245 two perpendicular number lines

B

Back-left view, 112 viewing an object from the left back corner

Back-right view, 112 viewing an object from the right back corner

Bar graphs, 62, 62 *illus.*, 64, 66, 68, 96 graphs that help to make comparisons among several items in a given category

Base, 98 In 10^{11}, the base is 10.

Base, 195, 219
area of, 193
hexagonal, 195

octagonal, 195
 radius of, 194-195
 triangular, 195
BASIC program
 compound interest, 263
 dart board simulation, 160
 estimate of area in the plane, 179-180
 handshake data, 28
 maximum area of a rectangle, 187
 random number generator, 140
 rolling two number cubes, 156
 sequences, 85
 timeline, 8
 to make a table of values, 218, 240
 to simulate a probability experiment, 147
Beats per minute, 224
Biased, 43, 48   distorted; chosen based on
 preference
Bimodal, 49   a sample that has two modes
Brainstorm, 92   a group discussion about possible
 ways to solve a problem

Calculator functions
 constant, 19
 d/c, 19
Calculators, 4, 20-22, 24, 28, 54, 138, 233
 constant function, **19**
 for graphing, 199
 order of operations, **16**
 pi, 189
 prime factorization, **17**
 problem solving, 18, 30
 reciprocal, **19**
 scientific, 16
 scientific notation, 98, 100
Calories, 25-26, 98, 100
Cartesian coordinate system, 22   also called
 rectangular coordinate system; a graphing method
 using the intersection of two perpendicular
 number lines that intersect at point zero
Celsius, 232
Center, 188, 189, 192

Centimeters, 61, 86, 127, 168, 190, 193, 196, 199,
 200, 214, 215, 234, 242, 243
Chance, 130-163
 how occurs, 136-137
 experiments and games of, 144-147
 experiments with, 152-156
 with circular spinners, 157-160
Circle, 157-159, 189 *illus.*, 193-196
 area of, 188, 190 *illus.*, 190-192, 195
 circumference of, 188-190
 properties of, 188-192
 radius of, 188, 190-191
 used to inscribe, 198
Circumference, 89, **188**-192, 195, 226   the
 perimeter of a circle
Coefficient, 230, 237   a multiplier of x; in the
 algebraic equation $y = ax + b$, a is the multiplier
Combination, 149, 184
Compound interest, 263   interest paid on the
 interest already earned as well as the principal
Compounding, 262   the process whereby earned
 interest is credited to your account so that
 this amount also begins to earn interest
Computer functions
 RND, 140, 160
Computers, 16, 19, 233, 240
 to generate random numbers, 136, 147
 to simulate situations, 132
 to solve equations, 217-218
Computer Activities
 area in a plane, 179-180
 compound interest, 263
 dart throwing, 160
 data base, **46,** 53, 55, 57
 graphing equations, 199, 200, 201
 handshake data, 28
 maximum area of a rectangle, 187
 mean, median, mode, 53
 random number generator, 140, 147
 rolling a number cube, 156
 sequences, 85
 spreadsheet, **18,** 46, 53, 55, 57, 218
 table of values, 218, 240
 timeline, 8
Computer spreadsheet, 19-20, 218
Concentric circles, 159 *illus.*
Confidence level, 66   tells about the precision
 of the survey along with margin of error

E

Edge, 14, **106,** 106 *illus.*   a segment where two faces of a polyhedron intersect
of cube, 113

Ellipse, 200, 201

Endpoints, 171, 197

Envelope, 199, 240-241   the shape or outline of a special curve
of a hyperbola, 200-201
of a parabola, 199
of an ellipse, 200

Equations, 206, 209, 217-223, 230-231
linear, 232-240
nonlinear, 240-244
solving, 236-238

Equilateral triangle, 172

Estimates, 83, 88, 91, 93, 157, 159, 167, 191, 192, 220, 236   a close approximation of the value, amount, size, weight, of something; based upon a strategy and calculation rather than a random guess
of area in the plane, 179-189
using proportional reasoning, 86

Even number, 142, 143, 155, 158, 162, 242

Expected value, 180

Experimental probability, 141   the number of times a particular item is selected compared to the total number of trials; also called relative frequency

Experiments, 139
analysis of, 144-147, 214-215
probability, 141-147
with chance, 152-156

Exponent, 98   In 10^{11} , the exponent is 11.

Exponential growth, 274   rapid growth in the life of a population

Expressions, 206
exponential, 264, 269, 271

Extrapolate, 220, 257

F

Face, 14, **106,** 106 *illus.*   the flat surface of a polyhedron formed by polygons and their interiors
of cube, **112,** 124, 162

Fahrenheit, 232

Feet, 86, 128, 186, 187, 190, 191, 192, 203

Formulas, 166, 206, 208, 215   rules that state the relationship between certain measurements
for area of circles, 191
for area of parallelograms, 178
for area of rectangles, 176
for area of trapezoids, 179
for area of triangles, 178
for circumference, 190

Foundation drawings, 115-120, 115 *illus.*, 123, 129   drawings that show the shape of the foundation placement, and the number of cubes that are built on this foundation

Fractions, 19, 30, 157

Frequency table, 41, **47,** 47 *illus.*, 48   a chart that indicates the number of values in each interval

Front view, 118, 119   the view seen looking directly at the front of an object rather than viewing the object from a corner; shows the length and height of the object

Front-left view, 112   viewing an object from the left front corner

Front-right view, 112   viewing an object from the right front corner

Function, 204-222, **223,** 224-245   a set of ordered pairs that has, at most, one value for the dependent variable for each value of the independent variable
linear, 230-235
nonlinear, 239-244

Function relation, 223

G

Gallons, 98, 128, 167, 227, 228, 235

Game
  creating, 163
  data, 148-152, 158-161
  fairness, 149-151
  of chance, 144-147
  number cube, 152-154
  patterns, 10
  strategies, 10, 192

Games
  Build It!, 14-15
  Pico, Fermi, Bagels, 11
  Rainbow Logic, 9-10

Geoboard, 172, 184, 197-198

Geometric
  curve, 214, 234
  drawings, 168
  ideas, 167-169, 203
  pattern, 169
  shapes, 4, 112, 170, 172, 188, 191

Geometric mean, 275 the amount in the middle of
  an interval of exponential growth; the square
  root of the product of the amount at the
  beginning of growth and at the end

Geometry, 168, 202
  adjacent, 9
  angle bisector, **171**
  angles, 170-171, 172
  area, 92, 92, 99, 110, 166, 169, 174-187, 190-192
  base, 193-195, 219
  circle, 157-159, 188, 189 *illus.*, 190-196
  circumference, 89, **188**-192, 195, 226
  compass, 188
  diameter, **188**, 189, 190, 196, 214, 215, 235
  dimensions, **93**, 94, 113, 122, 124, 186, 187,
    194-196
  edge, 14, 106-113
  endpoints, 171
  face, 14, 106, 112, 124, 162
  height, 94, 109, 175, 191, 193, 194, 195, 219, 226
  intersecting lines, 171
  length, 94, 96, 113, 181, 184, 188, 214, 215, 216,
    226
  line, 171, 217-220, 230

line segment, 113, 171, 188, 189, 197, 200, 214,
  234
midpoint, 171
parallelogram, 171, 173, 175-179, 191
perimeter, 166, 174, 181-187
perpendicular, 60, 171, 172, 199
point, 138, 161, 171, 180, 188, 189, 199, 200,
  201, 210, 214, 220, 228, 230, 235, 236, 241
polygon, 169, 170, 172-174, 179-180, 182-183,
  198
protractor, 107, 157-158, 171, 172
quadrilateral, 172, 173
radius, **188**, 190-191, 194-195, 226
rectangle, 171, 173, 174, 176, 177, 184, 186, 191,
  195, 198, 203
rhombus, 171, 173, 216
scale drawing, 86, 93, 94, 102, 107, 167-169
square, 127, 151, 160-161, 171, 173, 180, 181-
  182, 188-189, 192
spatial, 14, 108, 111
straight line, 188
surface area, 107, 109, 166, 168-169, 125-127,
  129, 195-196
three-dimensional, 12, 14, 106-107, 109, 193-196,
  111-114, 118-123
trapezoid, 171, 173, 178-179
triangle, 171-172, 175-178, 216
two-dimensional, 22, 60, 106
volume, 92, 93, 94, 99, 125, 166
width, 94, 175, 216

GNP, 56, 80

Grams, 25-26, 101

Graphical representation
  frequency table, 41, **47**, *illus.*
  isometric drawings, 107, **111**, 115-116,
    116-117 *illus.*, 122-123
  line graphs, **62**, 63
  line plots, **54**, 55, 58
  number line, 4, 8, 20, **22**, 29, 138-139
  of data, 62-64, 66-68, 184
  orthographic projections, **118**-123, 129
  parabola, **199**, 200
  rectangular coordinate system, **22**
  scale drawings, **86**, 93, 94, 102, 107, 167-169
  scatter plots, **60**, 63
  stem-and-leaf plot, **56**-59
  three-dimensional, **14**, 106-107, 109, 193-196
  timeline, **5**, 6-7, 22-23, 32
  tree diagram, 141, **144**-146, 146 *illus.*, 147, 148-
    150, 155, 159

H

Half life, 270   the length of time it takes for an organic substance to disintegrate to half it's original value

Headwind, 222   occurs when the wind is coming toward the front of the plane

Hidden lines, 119   dashed lines used to show surfaces that cannot be seen from the front view

Horizontal, 22   left to right

Hyperbola, 201   two curves that are the reflection image of each other and equidistant from two fixed points

I

Intervals, 57, 60 *illus.*, 62   the difference between successive values on a scale

Isometric drawings, 107, **111**, 115-116, 116-117 *illus.*, 122-123   drawings that show the corner view and the top or bottom of a three-dimensional figure

Isometric grid, 112   the arrangement of dots that can be used to draw a cube Isometric view, 115-117, 129

GLOSSARY/INDEX

Key, 56 explains the meaning of given information
Kilograms, 30, 94
Kilometers, 27, 89, 172, 190, 244

Lag phase, 274 slow growth in the early life of a population
Leaves, 56 in a stem-and-leaf plot, the digit(s) in the greatest place value(s)
Length, 94, 96, 113, 181, 184, 188, 214, 215, 216, 226
Line graphs, 62, 63 graphs used to show the behavior of a variable
Line plot, 54, 55, 58 statistical data displayed on a number line with marks above each datatype to indicate the number of pieces of data in each category
Line segment, 189, 197, 200, 214, 234
   diameter, 188
   edge, 113
   model, 170
   parallel, 171
   radius, 188
Linear equations, 232-235
   solving, 236-240
Linear functions, 230-235
   properties of, 230-232
Lines, 170, 171, 199, 217-220, 230
   graphs of, 217-219
   horizontal, 230, 245
   intersecting, 171
   steepness of, 216, 231, 245
   vertical, 230, 245
Logic, 11, 12, 14 the study of formal reasoning
   puzzle, 11

Margin of error, 66 tells about the precision of the survey along with confidence level
Matzeliger, Jan Earnst (1852-1889), 105
Maximum, 169, 185, 228, 232
   area, 186-187, 191
   perimeter, 184
Mean, 49, **50**-59, 96, 168 the balance point of a set of data; average
Measurement, 106, 125, 191, 202
   aerobic heart rate, 26
   area, 92, 93, 99, 110, 166, 168-169, 172, 174-188, 190-193, 195, 202-203
   beats per minute, 224
   calories, 25-26, 98, 100
   centimeters, 61, 86, 127, 168, 190, 193, 196, 199, 200, 214, 215, 234, 242, 243
   circumference, 89, 188-192, 195
   cubic centimeters, 93, 94, 101, **125,** 193, 234, 235
   cubic feet, 109
   cubic inches, 93
   days, 84, 85, 109, 234, 235
   decimeters, 101
   degrees, 157, 159, 227, 233
   f-stops, 269-270
   feet, 86, 128, 186, 187, 190, 191, 192, 203
   gallons, 98, 128, 167, 227, 228, 235
   GNP, 56, 80
   grams, 25-26, 101
   heart rate, 26
   hours, 27, 45, 84, 229, 234, 235, 244
   inches, 190, 191, 193, 194, 195
   ingots, 108
   kilograms, 30, 94
   kilometers, 27, 89, 172, 190, 244
   lung capacity, 109
   meters, 94, 101, 214, 215, 234, 236, 242, 243
   miles, 221, 227, 232
   miles per hour, 222, 228
   miles per second, 221
   milligrams, 101
   millimeters, 226, 235
   minutes, 27, 44, 64, 102, 103, 229, 235, 244
   months, 84
   ounces, 100, 235

GLOSSARY/INDEX

perimeter, 166, 168-169, 174, 181-187
population density, 95, 96
pounds, 223
pulse rate, 109
rate of usage, 169
revolutions, 190
seconds, 84, 101, 102, 236
square centimeters, 93, 100, **125**
square feet, 109, 128, 184, 186, 235
square inches, 93, 193
square kilometers, 95, 101
square meters, 94
square miles, 95
strides, 87-89
surface area, 107, 109, **125**, 126, 129, 166-169, 195-196
tablespoons, 273
temperature, 23, 109, 232
time, 5-6
volume, 92, 93, 94, 99, 107, 109, **125**, 126, 128, 129, 166, 168, 193-196, 242
weeks, 84, 103
weight, 25-26, 100, 101, 224, 227
years, 84, 85, 96, 103, 109, 135, 221, 233, 234

Measures of central tendency, 41, **49**-55, 96   three common measures that help describe a set of data; mode, median, mean
mean, 49, **50**-59, 96, 168
median, **49**-59, 96, 168
mode, **49**-59, 96, 169

Median, 49-59, 96, 168   middle of the data; when the data are arranged in order, the point where exactly half the values are greater and half the values are less

Meters, 94, 101, 214, 215, 234, 236, 242, 243

Midpoint, 171

Miles, 221, 227, 232

Milligrams, 101

Millimeters, 226, 235

Minutes, 27, 44, 64, 102, 103, 229, 235, 244

Minimum, 169, 185
   area, 186
   perimeter, 184

Mixed numbers, 19

Mode, 49-59, 96, 169   the most frequent value in a set of data

Model
   for predicting, 144

for developing a formula, 220
   of a line segment, 170
   of a triomino, 185
   with the ellipse, 201
   with the hyperbola, 201
   with the parabola, 201
Months, 84
Morgan, Garrett A. (1877-1963), 3
Morita, Akio, 131
Multiple, 142

N-gon, 173   a polygon with *n* sides

*n*th term, 210-216, 245

Nesmith, Bette, 77

Net, 125, 126, 128, 129, 195 *illus.*   the flat pattern that would fold up with no gaps or overlaps to form a cube

Nonlinear equations, 240-244

Nonlinear functions, 239-244

Number line, 4, 8, 20, **22**, 29, 138-139   a line that has a number assigned to each point on it

Number sense, 78-79

Numbers
   decimal, 19, 20, 30, 190
   difference, 153-154, 155, 162, 222
   divide, 158, 191
   even, 142, 143, 155, 158, 162, 242
   exploring, 16-21
   fractions, 19-20, 30, 157
   mixed, 19
   multiple of, 142
   odd, **21,** 142, 143, 153-154, 155, 158, 162, 185, 242
   ordered pairs of, **22**-25, 27
   patterns, 4, 122, 124, 198
   percents, **47,** 142, 144
   pi, 30, 189
   powers, 98
   prime, 142, 154, 242
   prime factorization of, **17**
   product, 153-154, 162, 242

random, 132, 136, 140, 161, 162
real, 217
reciprocal, **19**, 30
scientific notation, **97**, 98, 99-101
sequences, 17, 19, 122
sets of, 17, 22
square root, 17
squares of, 30
standard form, **97**, 98
sum, 30, 152, 153, 154, 155-156, 158, 198, 222
whole, 162
Numerical data
sequencing, 4, 19
Numerical scale, 5

O

Object lines, 119   solid lines that show surfaces that can be seen when looking from a particular view
Obtuse triangle, 172, 177 *illus.*
Octagon, 172, 193, 195, 198
Odd number, 21, 142, 143, 153-154, 155, 158, 162, 185, 242   a whole number that is not even
Odds, 152, **154**, 155   the ratio of the likelihood an event will happen to the likelihood it will not happen
Opposite
pairs of angles, 171
parallel sides, 172
Order of operations, 16   the rules to follow when more than one operation is used 1. Do all operations within grouping symbols first; start with the innermost grouping symbols. 2. Do all powers before other operations. 3. Do multiplication and division in order from left to right. Do all addition and subtraction in order from left to right.
on calculators, 16
Ordered pair, 22-25, 27, 151, 221-244   a pair of numbers where order is important; an ordered pair that is graphed on a coordinate plane is written in the form: (*x*-coordinate, *y*-coordinate)

Ordered quadruples, 147
Ordered triple, 146   a triple number of picks or draws where order is important
Ordering, 4, 146, 147
Organizing, 4
data, 47-48
Origin, 22   zero on the number line; point of intersection of *x*-axis and *y*-axis in a coordinate system
Orthographic projections, 118-123, 129   multiview drawings which enable a drafter to accurately describe on paper the size and shape of any object
Ounces, 100, 235
Outcomes, 148, 150, 157
favorable, 137
possible, 137, 145, 146, 149, 152
probability of, 138
random, 148
Output, 187

P

Paper Folding, 199-201, 240-241
Parabola, 199, 200   a curve that is the same distance from a fixed point and a fixed line
Parallel, 172
Parallelogram, 171, 173, 175 *illus.*, 191
area, 175-179
Pascal, Blaise (1623-1662), 198
Pascal's Triangle, 198
Patterns, 29, 61, 82, 127, 182, 245
extending, 210-211, 239-240
for all points of a hyperbola, 201
for all points of an ellipse, 200
for problem solving, 212-216
in Pascal's Triangle, 198
of cubes, 107, 122-125
of data, 28, 173, 184, 199, 206
of decimals, 20
of figures, 211, *illus.*
of games, 10
of geometric shapes, 4, 169

GLOSSARY/INDEX

Scatter plots, **60,** 61 *illus.*, 63   useful for seeing if a relationship exists between two variables with patterns

Scientific calculators, **16**   calculators that follow the order of operations rule

Scientific notation, **97,** 98-101   a way of writing numbers as the product of a number that is at least 1 but less than 10 and a power of 10

Seconds, 84, 101, 102, 236

Segment bisector, 171

Sequencing, 10, 151, 160
  BASIC program, 85
  calculator operations, 17, 18
  events, 5
  numerical data, 4, 17, 19, 122

Sets, 17
  of numbers, 17, 22
  using Venn diagrams, 12

Sieve of Erasthosenes, 17

Simple interest, 262

Simulate, 132
  probability experiment, 147
  rolling a number cube, 156

Soma cube, **122,** 126   a 3-by-3-by-3 cube

Spatial, **14,** 108, 111   relating to space
  reasoning, 129

Spatial figures
  for problem solving, 129

Spatial visualization, 104-129

Spreadsheet, **18,** 28, 46, 53, 55, 57, 218   a computer tool used for organizing and analyzing data and formulas, which are arranged in row and column format

Square, 127, 151, 160-161, 171, 173, 180, 181, 188-189, 192, 198
  area of, 182, 183
  to make a triomino, 185
  to make dominoes, 185
  to make hexominoes, 186
  to make pentominoes, 186
  to make tetrominoes, 186

Square centimeters, 93, 100, **125**   the area measurement using centimeters

Square feet, 109, 128, 184, 186, 235

Square inches, 93, 193

Square kilometers, 95, 101

Square meters, 94

Square miles, 95

Square root, 17

Square root function, 242

Squares of numbers, 30

Standard form, **97,** 98   the standard form for one hundred twenty-five is 125

Stationary phase, **274**   slower growth in the life of a population

Statistics, 54, 71, 82
  bar graphs, 62
  charts, 27, 39, 48
  data, 45-75
  data bases, **46,** 53, 55, 57
  frequency tables, 41, **47,** 47 *illus.*, 48
  graphs, 65-68
  line graphs, 62-63
  mean, **50**-59, 96
  measures of central tendency, **49**-55, 96
  median, **49**-59, 96
  mode, **49**-59, 96
  range, **5,** 46, 60-62
  samples, 69-73
  scatter plots, **60,** 61, 63
  spreadsheets, **18,** 28, 46, 53, 55, 57
  stem-and-leaf plots, **56**-59
  tables, 28, 63-64

Stem-and-leaf plot, **56**-59   a graph where the digits in the greatest place values of the data values are the stems and the digits in the next greatest place values are the leaves

Stem, **56**   in a stem-and-leaf plot, the digits in the second greatest place values

Strategy, 9-13, 83-84
  game, 10, 152-153, 192

Sum, 30, 155, 158, 198, 222
  of angles, 172
  odd, 154
  of number cubes, 152-153, 156
  patterns, 198
  probability of, 159

Surface area, **125,** 166   the number of unit squares needed to cover the outside of a three-dimensional object
  body, 109
  building, 129
  containers, 168-169
  cube, 126

cylinders, 195-196
dome, 109
prism, 107, 127
Surface edges, 118
Survey, 27, 38, **42,** 44-48, 69 *illus.*   information
gathered through questioning
data, 42-**43,** 138
designing, 39
evaluating, 41, 65-66, 69-73
planning, 40-41
questionnaire, 43, 45-47, 64
sample, **43**

Table, 28, 63-64, 181-182, 181 *illus.*, 189
frequency, 41, 47 *illus.*, 48
of data, 224 *illus.*
Tablespoons, 273
Tailwind, 222   occurs when the wind is coming
toward the tail of the plane
Tally, 47-48, 138, 142
Technology, see Calculators and Computers
Temperature, 23, 109, 232
Term, 210-216, 239
Tetrominoes, 186
Theoretical probability, 132, **141,** 142, 143, 149,
153   the ratio of the number of elements in the
event compared to the number of elements in
the sample space
Theories, 220, 230, 231
Three-dimensional, 14, 106-107, 109, 193-196
denotes three measurements enclosing a part of
space
isometric drawings, 111-114
orthographic projections, 118-123
Time, 5-6
Timeline, 5, 6-7, 22-23, 32   a kind of number line
that can be used to help interpret past events
number lines, 8
Top view, 118-119   the view seen looking directly
at the top of an object rather than viewing the
object from a corner; projected directly above
the front view showing the

length and depth
Torres, Antonio de, 165
Trapezoid, 171, 173
area, 178-179, 178 *illus.*
Tree diagram, 141, **144**-146, 146 *illus.*, 147, 148-
150, 155, 159   a diagram used to organize
information to show the total number of
possible outcomes in a probability experiment
Trend, 60, 220   the result of data or tendency of
the data
Triangle, 171, 198, 216
acute, 171-172, 177
area, 175-178, 219
equilateral, 172
isosceles, 172
obtuse, 172, 177 *illus.*
right, 172, 175 *illus.*, 177
scalene, 172
vertices, 177
Trimodal, 49   a sample that has three modes
Triomino, 185-186   a pattern made by joining
exactly three squares together so they share sides
Two-dimensional, 22, **60,** 106   denotes two
measurements

Unbiased, 43   chosen randomly so there is no
over representation of one part of the sample
population
Unique, 15   has only one solution
Union, 197
Unit pricing, 235

V

Variables, 60, 206, 215   symbols representing
values
dependent variables, 210-212, 214-215, 220-224

Glossary/Index

independent variables, 211-212, 214-215, 220-224

used to solve problems, 220-222

Venn diagram, 12, 30   use of overlapping circles to show which items are shared

both, or, and, and not, 13

of quadrilaterals, 173

to organize sets, 12

Vertex, 106, 106 *illus.,* 227   a point where three or more edges intersect

of cube, **112**

Vertical, 22, 214   up and down

Vertical axis, 22, 62, 189, 210-211

Vertical line, 230

Vertices, 112, 117, 177, 188-189, 198

Volume, 92, **93,** 94, 99, **125,** 166, 242   the amount of space occupied by an object; the number of unit blocks needed to fill the interior space of an object

buildings, 129

containers, 168

cubes, 126

cylinders, 193-196

lungs, 109

prisms, 107, 128, 193-195

x-axis, 60, 62, 211, 241

y-axis, 60, 62, 211, 241

Years, 84, 85, 96, 103, 109, 135, 221, 233, 234

Zero, 22, 136

Wakefield, Ruth Graves, 205

Wealth, 33, 39, 80, 108-109, 133-134, 167-168, 207-208, 252

Weeks, 84, 103

Weight, 25-26, 100, 101, 224, 227

containers, 168

Whole numbers, 162

GLOSSARY/INDEX

PHOTO CREDITS

232, Todd Powell/Profilies West; 233, Todd Yarrington; 234, Michel Tchenevhoff/The Image Bank; 236, H. Wendler/The Image Bank; 237 (l), Matt Meadows, (r), Mark Madden; 241, Darryl Torckler/Tony Stone Worldwide/Chicago LTD.; 243, John Kelly/The Image Bank; 244, John Kelly/The Image Bank; 245, Springer-Verlag, New York; 246, Patrick Doherty/The Image Bank; 248,249, Courtesy Lockhed Sanders, Photo by Todd Yarrington; 250 (inset), David De Lossy/The Image Bank, ©Kunh, Inc./The Image Bank; 251, Steve Proehl/The Image Bank; 252 (inset), ©Berenholtz/The Stock Market, Glen Allison/Tony Stone Worldwide/Chicago LTD,; 253 (l), ©Stan Flint/The Image Bank, (r), Jeff Blanton/Tony Stone Worldwide/Chicago LTD.; 254, Todd Yarrington; 256, Aaron Haupt; 257, Duomo/Paul J. Sutton; 258, Joe DiChello Jr.; 260, David Madison; 261, Studiohio; 262, Greg Murphy/Tony Stone Worldwide/Chicago LTD.; 264, Todd Yarrington; 266, Cindy Lewis; 267, Pete Saloutos/Tony Stone Worldwide/Chicago LTD.; 268, Matt Meadows; 270, Tim Courlas; 272, David Weintaub/Photo Researchers; 273, Don Landwehle/The Image Bank; 274, Julie Houck/Tony Stone Worldwide/Chicago LTD.; 275, Kean Collection/Archive Photos; 276, Steve Elmore/The Stock Market; 277, Bruce Iverson; 278 (inset), Rohan/Tony Stone Worldwide/Chicago LTD., Archive Photos/Martin Forstenzer; 279, Todd Yarrington; 285 (bl) Sharon Remmen; 285 (br) Dan Nieman; 285 (c) Roger Burnard; 285 (tr) File Photo; 286, United Press International; 287 NASA.